Tobias Eltze

Consequences of altered poly(ADP-ribose) polymerase-1 expression

AF060796

Tobias Eltze

Consequences of altered poly(ADP-ribose) polymerase-1 expression

Effects of altered PARP-1 expression

Südwestdeutscher Verlag für Hochschulschriften

Impressum/Imprint (nur für Deutschland/only for Germany)
Bibliografische Information der Deutschen Nationalbibliothek: Die Deutsche Nationalbibliothek verzeichnet diese Publikation in der Deutschen Nationalbibliografie; detaillierte bibliografische Daten sind im Internet über http://dnb.d-nb.de abrufbar.
Alle in diesem Buch genannten Marken und Produktnamen unterliegen warenzeichen-, marken- oder patentrechtlichem Schutz bzw. sind Warenzeichen oder eingetragene Warenzeichen der jeweiligen Inhaber. Die Wiedergabe von Marken, Produktnamen, Gebrauchsnamen, Handelsnamen, Warenbezeichnungen u.s.w. in diesem Werk berechtigt auch ohne besondere Kennzeichnung nicht zu der Annahme, dass solche Namen im Sinne der Warenzeichen- und Markenschutzgesetzgebung als frei zu betrachten wären und daher von jedermann benutzt werden dürften.

Coverbild: www.ingimage.com

Verlag: Südwestdeutscher Verlag für Hochschulschriften GmbH & Co. KG
Heinrich-Böcking-Str. 6-8, 66121 Saarbrücken, Deutschland
Telefon +49 681 37 20 271-1, Telefax +49 681 37 20 271-0
Email: info@svh-verlag.de

Approved by: Konstanz, Universität Konstanz, Diss. , 2009

Herstellung in Deutschland:
Schaltungsdienst Lange o.H.G., Berlin
Books on Demand GmbH, Norderstedt
Reha GmbH, Saarbrücken
Amazon Distribution GmbH, Leipzig
ISBN: 978-3-8381-3107-8

Imprint (only for USA, GB)
Bibliographic information published by the Deutsche Nationalbibliothek: The Deutsche Nationalbibliothek lists this publication in the Deutsche Nationalbibliografie; detailed bibliographic data are available in the Internet at http://dnb.d-nb.de.
Any brand names and product names mentioned in this book are subject to trademark, brand or patent protection and are trademarks or registered trademarks of their respective holders. The use of brand names, product names, common names, trade names, product descriptions etc. even without a particular marking in this works is in no way to be construed to mean that such names may be regarded as unrestricted in respect of trademark and brand protection legislation and could thus be used by anyone.

Cover image: www.ingimage.com

Publisher: Südwestdeutscher Verlag für Hochschulschriften GmbH & Co. KG
Heinrich-Böcking-Str. 6-8, 66121 Saarbrücken, Germany
Phone +49 681 37 20 271-1, Fax +49 681 37 20 271-0
Email: info@svh-verlag.de

Printed in the U.S.A.
Printed in the U.K. by (see last page)
ISBN: 978-3-8381-3107-8

Copyright © 2012 by the author and Südwestdeutscher Verlag für Hochschulschriften GmbH & Co. KG and licensors
All rights reserved. Saarbrücken 2012

Summary

Poly(ADP-ribosyl)ation is a posttranslational modification of cellular proteins mostly catalysed by poly(ADP-ribose) polymerase-1 (PARP-1) and to a lesser extent by PARP-2. PARP-1 and PARP-2 use NAD^+ as substrate in response to cellular exposure to various DNA-damaging agents to form the biopolymer poly(ADP-ribose) (PAR). On the one hand, poly(ADP-ribosyl)ation is a key regulator of genomic stability under conditions of genotoxic stress where PARP-1 plays a major role in DNA repair, transcription regulation and recovery of cells after DNA damage. On the other hand, massive poly(ADP-ribosyl)ation induced by severe acute DNA damage results in rapid depletion of cellular NAD^+ and ATP pools, which can lead to cell death.

In three sub-projects, this study aimed to explore the role of PARP-1 in DNA repair and cell death using *in vitro* as well as *in vivo* models:

(i) PARP-1 is involved in a number of pathophysiological conditions such as diabetes, inflammation and stroke, consequently, pharmacological inhibitors of PARP have the potential to elicit beneficial effects in these diseases. In the first part of the present study, a new PARP inhibitor, BYK204165, was examined for inhibition of PAR-synthesis in H_2O_2-treated 3T3 fibroblasts from *Parp-1*$^{+/+}$ and *Parp-1*$^{-/-}$ mice, where the 100-fold PARP-1 selectivity of the compound was confirmed by its failure to inhibit PARP-2 in both cell lines. The new compound might provide a novel and convenient functional tool toward the assessment of the contribution of PARP-1 and PARP-2 related PAR formation in intact cells, because the enzymatic activity of the two isoforms can be distinguished by use of a selective PARP-1 inhibitor.

(ii) Since inhibition of PAR formation generally influences DNA repair mechanisms, the second part of the study explored the consequences of stably overexpressed human PARP-1 (hPARP-1) in Chinese hamster cells (COMF10) on the cytotoxicity induced by alkylating agents (MMS, MNNG) and X-irradiation. Measurements of apoptosis, necrosis, DNA repair and genomic stability were taken as experimental endpoints. Analysis of cell viability after treatment with MMS and MNNG revealed consistently larger fractions of necrotic cells in the COMF10 cells compared to control. Furthermore, DNA repair kinetic measurements after X-irradiation of hPARP-1 overexpressing murine lymphoma EL-4 cells, demonstrated acceleration in DNA repair, whereas pharmacological inhibition of PARP by PJ34 delayed and reduced DNA repair capacity.

(iii) Finally, it was intended to generate an *in vivo* system for tissue-specific overexpression of hPARP-1 protein in mice. Therefore, 18 transgenic founder mice were generated by DNA microinjection with a transgene comprising *hPARP-1* cDNA under the control of a strong promoter.

Summary

The transcription is "locked" by a Stop-sequence that can be eliminated *in vivo* by expression of Cre recombinase as a result of crossing with appropriate transgenic tissue-specific "Cre-deleter" mice (at first T-cell specific). Unexpectedly however, although mRNA transcripts of the *hPARP-1* transgene could be found in their offspring, for largely unknown reasons its respective protein expression could not be detected. Its failure on the level of translation possibly could be due to unexpected transcriptional start sites within the transgene. Therefore, an alternative approach should be used for follow-up projects in order to obtain hPARP-1-overexpressing mice.

In summary, studies within the work of this thesis contributed to the disposal of a novel and selective PARP-1 inhibitor, which provides a valuable tool to dissect different roles of PARP-1 and PARP-2 in cellular functions (Eltze *et al.*, 2008). Moreover, it was shown that overexpression of hPARP-1 in rodent cells has two important consequences. Its overexpression leads to an increased DNA repair capacity after X-irradiation, but on the other hand also to an increased susceptibility to DNA damage in response to alkylating agents or PARP inhibition, demonstrating the dual role of PARP-1 in mechanisms of DNA repair and cell death (Eltze and Kunzmann *et al.*, submitted). Finally, novel transgenic mice with intended tissue-specific overexpression of hPARP-1 were generated and characterized on a genetic level. Unexpectedly, despite cell culture validation of the expression construct and transgene expression on the mRNA level, no protein expression of the transgene could be detected in these mice for largely unknown reasons. This outcome needs to be considered in future approaches aiming at the generation of *hPARP-1* transgenic mice.

TABLE OF CONTENT

1 INTRODUCTION 7

    1.1 General role of PARP   7
    1.2 Structure of PARP-1   8
    1.3 NAD^+ metabolism   9
    1.4 Poly(ADP-ribose) metabolism   10
    1.5 Regulation of PARP activity   11
    1.6 Role of PARP-1 in modulation of chromatin structure   13
    1.7 DNA repair   14
        1.7.1 Base excision repair (BER) and involvement of PARP-1/PARP-2   *14*
        1.7.2 Nucleotide excision repair (NER)   *15*
        1.7.3 Double-strand break (DSB) repair   *16*
    1.8 Implication of PARP in aging, longevity and genomic stability   17
    1.9 PARP and its implication in T-cell development   19
    1.10 The role of PARP-1 in cell survival and cell death   20
    1.11 PARP-2   23
    1.12 PARP-1 and PARP-2 mediated functions in knock-out mice   24
    1.13 Pharmacological inhibition of PARP   25

2 OBJECTIVE 29

3 MATERIAL AND METHODS 31

    3.1 Material   31
        3.1.1 Chemicals and reagents   *31*
        3.1.2 Laboratory equipment   *34*
        3.1.3 Buffers and solutions   *35*
        3.1.4 Plasmids   *37*
        3.1.5 Oligonucleotides   *38*
        3.1.6 PCR conditions   *40*
        3.1.7 Molecular weight standards   *40*
        3.1.8 Kits   *41*
        3.1.9 Enzymes   *41*
        3.1.10 Restriction enzymes   *42*
        3.1.11 Polymerases   *42*
        3.1.12 Antibodies   *43*

3.1.13 Cell lines	*43*
3.1.14 Organisms	*44*
3.1.15 Software	*44*
3.2 Methods	45
3.2.1 DNA repair measurement	*45*
3.2.1.1 Measurement of DNA strand breaks with FADU	45
3.2.1.2 Automated FADU protocol	46
3.2.2 DNA/RNA/protein isolation	*46*
3.2.2.1 Organ isolation from mice	46
3.2.2.2 DNA isolation from organs	47
3.2.2.3 RNA isolation from organs	47
3.2.2.4 Protein isolation from organs	47
3.2.3 Isolation and purification of T-cells from thymus and spleen	*48*
3.2.4 Genotyping of mice	*48*
3.2.5 Determination of transgene copy number	*48*
3.2.6 Cell culture	*49*
3.2.6.1 Cell number determination	49
3.2.6.2 Thawing of cells	49
3.2.6.3 Cryopreservation of cells	50
3.2.6.4 Cell culture and passaging of cells	50
3.2.6.5 Transient transfection of EL-4 cells with jetPEI™	50
3.2.7 Molecular biological methods	*50*
3.2.7.1 Restriction analysis	50
3.2.7.2 Agarose gel electrophoresis	51
3.2.7.3 Gel extraction	51
3.2.7.4 Phenol-chloroform extraction of DNA	51
3.2.7.5 Blunting and dephosphorylation of DNA fragments	52
3.2.7.6 Purification of reaction mixtures	52
3.2.7.7 Ligation	52
3.2.8 Preparation of chemo-competent E. coli DH5α	*52*
3.2.9 Transformation	*53*
3.2.10 Colony screening	*53*
3.2.11 Generation of hPARP-1 transgenic mice by DNA microinjection	*53*
3.2.11.1 Preparation of the transgene for DNA microinjection	53
3.2.11.2 Embryo isolation from mice	54
3.2.11.3 DNA injection	54
3.2.11.4 Embryo transfer in foster females	54
3.2.12 Protein characterization by SDS-PAGE	*55*
3.2.13 Western blot	*56*
3.2.14 Immunofluorescence	*56*
3.2.14.1 Analysis of *hPARP-1* from adherent cells	56

Table of content

3.2.14.2 Analysis of PAR in suspension cells	57
3.2.14.3 Analysis of PAR in fibroblasts	57
3.2.14.4 Cytotoxicity assay (microscopy)	58
3.2.15 Cytotoxicity assay (FACS)	*58*

4 RESULTS   60

4.1 Inhibition of PARP-1/-2 in mouse *Parp-1$^{+/+}$* and *Parp-1$^{-/-}$* fibroblasts	60
4.1.1 Detection and selective inhibition of PAR formation in Parp-1$^{+/+}$ and Parp-1$^{-/-}$ fibroblasts	*60*
4.2 DNA repair and viability in hPARP-1 overexpressing rodent cells	64
4.2.1 Toxicity induced by alkylating agents in hPARP-1 overexpressing COMF10 cells	*64*
4.2.2 Expression of hPARP-1 in murine lymphoma EL-4 cells	*66*
4.2.3 DNA repair kinetics in EL-4 cells treated with jetPEI^{TM}	*67*
4.2.4 Induction of DNA strand breaks by X-irradiation in EL-4 cells	*68*
4.2.5 Determination of the optimal X-irradiation dose for DNA repair measurements	*69*
4.2.6 Repair kinetics of DNA strand breaks in EL-4 cells with hPARP-1 overexpression	*70*
4.2.7 Perturbation of PARP activity by PARP inhibition with PJ34 in EL-4 cells	*71*
4.2.8 Repair kinetics of DNA strand breaks in EL-4 cells after PARP inhibition	*72*
4.3 Experiments for generating hPARP-1 transgenic mice	73
4.3.1 Generation of the hPARP-1 transgene for DNA microinjection	*73*
4.3.2 Functional expression analysis of the hPARP-1 transgene in vitro	*78*
4.3.3 Detection of hPARP-1 transgenic founder mice by real-time PCR	*79*
4.3.4 Detection of hPARP-1 transgenic founder mice by conventional PCR	*81*
4.3.5 Analysis of hPARP-1 protein level in transgenic mice	*81*
4.3.6 Excision of the Neo/Stop sequence in genomic DNA of hPARP-1 transgenic mice	*82*
4.3.7 Analysis of founder mice for complete transgene integration	*85*
4.3.8 Analysis of hPARP-1 mRNA transcription in transgenic animals	*85*
4.3.9 Relative quantification of the transgene copy number	*87*

5 DISCUSSION   90

5.1 Inhibition of PARP-1/-2 in mouse *Parp-1$^{+/+}$* and *Parp-1$^{-/-}$* fibroblasts	90
5.1.1 Detection and selective inhibition of PAR formation in Parp-1$^{+/+}$ and Parp-1$^{-/-}$ mouse fibroblasts	*90*
5.2 DNA repair and viability in hPARP-1 overexpressing rodent cells	92
5.2.1 Toxicity induced by alkylating agents in hPARP-1 overexpressing COMF10 cells	*92*
5.2.2 Expression of hPARP-1 in murine lymphoma EL-4 cells	*93*
5.2.3 DNA repair kinetics in EL-4 cells treated with jetPEI^{TM}	*94*
5.2.4 Induction of DNA strand breaks by X-irradiation in EL-4 cells	*94*
5.2.5 Determination of the optimal X-irradiation dose for DNA repair measurements	*94*

5.2.6 Repair kinetics of DNA strand breaks in EL-4 cells with hPARP-1 expression   95
5.2.7 Perturbation of PARP activity by PARP inhibition with PJ34 in EL-4 cells   96
5.2.8 Repair kinetics of DNA strand breaks in EL-4 cells after PARP inhibition   96
5.3 Experiments for generating *hPARP-1* transgenic mice   97
5.3.1 Generation of hPARP-1 transgene for DNA microinjection and its functional expression analysis   97
5.3.2 Detection of hPARP-1 transgenic founder mice by real-time PCR   98
5.3.3 Detection of hPARP-1 transgenic founder mice by conventional PCR   98
5.3.4 Analysis of hPARP-1 protein level in transgenic mice   99
5.3.5 Excision of the Neo/Stop sequence in genomic DNA of hPARP-1 transgenic mice   100
5.3.6 Analysis of founder mice for complete transgene insertion   101
5.3.7 Analysis of hPARP-1 mRNA transcription in transgenic animals and their transgene copy number   101
5.3.8 Reasons for the lack of hPARP-1 expression in transgenic mice   102
5.4 General summary and discussion   105

6 REFERENCES   107

7 APPENDIX   120

7.1 Abbreviations   120
7.2 Figures   121
7.3 Tables   122

1 Introduction

1.1 General role of PARP

Poly(ADP-ribosyl)ation is a posttranslational modification of cellular proteins in eukaryotic cells catalysed by poly(ADP-ribose) polymerases (PARPs) using nicotinamide adenine dinucleotide (NAD^+) as substrate in response to cellular exposure to various DNA-damaging agents. The DNA-dependent formation of poly(ADP)ribose (PAR) in nuclear extracts from rat liver stimulated by NAD^+ was discovered by Chambon and colleagues more than forty years ago (Chambon et al., 1963). This observation initiated rapidly growing research in the field of poly(ADP-ribosyl)ation and succeeded in the discovery of various enzyme isoforms of PARP being involved in the regulation of many cellular functions. Today, the most extensively studied enzyme producing PAR is named poly(ADP-ribose) polymerase-1 (PARP-1), which is an abundant nuclear enzyme of the PARP protein superfamily. This enzyme family currently consists of 17 homologous genes that share a conserved catalytic domain but differ in their molecular structure and subcellular localization (Shall and de Murcia, 2000; Bürkle, 2005; Hassa et al., 2006; Schreiber et al., 2006). PARP-1 is able to polymerize linear or branched chains of ADP-ribose covalently attached to a variety of acceptor proteins, primarily PARP-1 itself, but also histones, topoisomerases, DNA polymerases, DNA ligases, several DNA repair factors, transcription factors and cell cycle factors (reviewed in (D'Amours et al., 1999). Meanwhile, poly(ADP-ribosyl)ation reactions are known to be involved in many cellular processes under physiological conditions like DNA repair pathways, cell survival, genomic stability and longevity (D'Amours et al., 1999; Bürkle et al., 2005). Moreover, in vitro experiments have shown that PARP-1 activity can regulate chromatin structure, as covalently bound PAR on histones can induce their release from the DNA. This leads to a suggested histone shuttle mechanism in vivo, where the induced chromatin relaxation by PARP-1 activity may guide specific proteins to sites of repair and promote DNA base excision repair (Althaus, 1992). However, in pathophysiological conditions, e.g. after generation of free radicals, reactive oxygen species and peroxynitrite, excessive poly(ADP-ribosyl)ation reactions consume NAD^+ and consequently ATP, culminating in cellular dysfunction by slowing the rate of glycolysis and mitochondrial respiration. The results of this process favor either the risk to promote apoptosis by the release of apoptosis-inducing factor (AIF) from mitochondria or necrosis as a consequence of NAD^+ and ATP depletion. PAR synthesis induced by PARP-1, but not by PARP-2, has been shown to promote translocation of AIF from mitochondria to the nucleus with subsequent DNA fragmentation, and thus to mediate a caspase-independent apoptotic pathway (Yu et al., 2002; Andrabi et al., 2006; Yu et al., 2006; Cohausz et al., 2008). Finally, PARP-1 participates in regulating inflammation, as it acts as a co-activator of the transcription factor nuclear factor-kappaB (NF-κB) resulting in the synthesis of pro-inflammatory mediators, or by direct poly(ADP-

Introduction

ribosyl)ation of transcription factors like STAT and activator protein-1 and -2 (Kauppinen, 2007). These deleterious mechanisms of PARP overactivation are implicated in several diseases such as ischemia in various organs (stroke, myocardial infarction), circulatory shock, diabetes, neurodegenerative disorders including Parkinson's and Alzheimer's disease, allergy, rhinitis and other inflammatory disorders (Pieper *et al.*, 1999b; Virág and Szabó, 2002; Iwashita *et al.*, 2004c; Chiarugi, 2005; Besson, 2009). Therefore, inhibition of PAR synthesis by appropriate and selective inhibitors constitutes an attractive approach for tissue protection, provided that the beneficial effects of PARP-1 to repair damaged DNA and to preserve nuclear integrity remain undisturbed (Besson, 2009).

1.2 Structure of PARP-1

Human PARP-1 is a 113-kDa enzyme (EC 2.4.2.30) and encoded by the *ADPRT* gene on chromosome 1 located at position 1q41-q42. The enzyme is composed of three major domains with distinct functions: the DNA binding domain (DBD) at the amino terminus (amino acid, aa1–aa373), the central automodification domain (aa374–aa525), and the catalytic domain (CAT) located at the carboxyl terminus (aa526–aa1014) (Figure 1) (D'Amours *et al.*, 1999). The DNA binding domain (46 kDa) consists of two zinc finger domains, ZFI (aa21-aa56) and ZFII (aa125-aa162), which are both necessary for binding to nicked DNA (de Murcia and Menissier de Murcia, 1994). DNA double-strand breaks (DSBs) are mainly recognized by zinc finger FI, whereas activation of PARP-1 by single-strand DNA breaks (SSBs) requires both zinc fingers (Ikejima *et al.*, 1990). In addition, a third zinc binding domain (aa216-aa366) with a zinc ribbon fold was recently discovered that could be necessary to relay the DNA binding signal from the first two zinc fingers to the catalytic carboxy terminus by forming an interdomain contact, which is important for DNA-dependent PARP-1 activation (Langelier *et al.*, 2008). Furthermore, the DBD contains a bipartite nuclear localization signal (NLS, aa207-aa226) for the nuclear homing of PARP-1 (Alvarez-Gonzalez *et al.*, 1999), and within the NLS, a caspase-3 and caspase-7 cleavage site (DEVD, aa210-aa213, Asp-Glu-Val-Asp), where PARP-1 is cleaved in the execution phase of apoptosis (Duriez and Shah, 1997). As a consequence, the cleavage of PARP-1 generates two proteolytic fragments, a 24-kDa amino terminus and an 85-kDa carboxyl terminus (Lazebnik *et al.*, 1994; Tewari *et al.*, 1995; Germain *et al.*, 1999). The automodification domain (16 kDa) is a central regulating segment comprising a breast cancer-susceptibility protein-carboxy (C) terminus motif (BRCT). The BRCT motif is common in many DNA-repair and cell-cycle proteins and allows protein-protein interactions. The C-terminal catalytic domain (55 kDa) accommodates the designated PARP signature sequence, a 50-amino acid sequence showing 100% homology between vertebrates (Virág and Szabó, 2002). This domain contains a NAD⁺ fold that polymerizes ADP-ribose units to linear or branched chains on respective target proteins. Interestingly, the PARP signature sequence shows common structural features with the active site of bacterial (ADP-ribosyl)ating toxins like diphteria toxin or pertussis toxin (Amè et al., 2004).

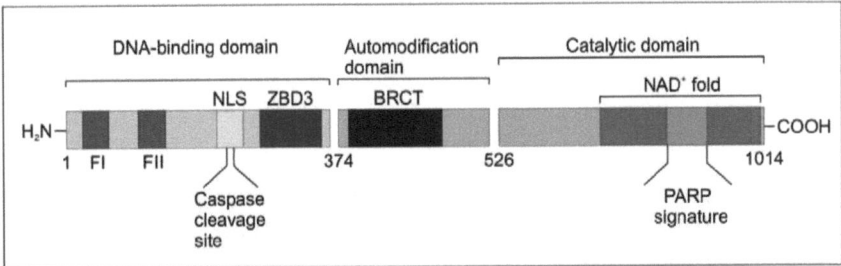

Figure 1: Schematic structure of the modular organization of human PARP-1. Human PARP-1 consists of three major structural and functional domains: (1) the DNA-binding domain contains two zinc fingers (FI, FII), a nuclear localization signal (NLS) comprising a caspase cleavage site and a third zinc binding domain (ZBD3), (2) the central automodification domain with a breast cancer-susceptibility protein-carboxy (C) terminus motif (BRCT), and (3) the catalytic domain with the NAD⁺ binding site (NAD⁺ fold) with the conserved PARP signature. The beginning of each domain is indicated by amino acid numbering.

1.3 NAD⁺ metabolism

In eukaryotic cells, NAD⁺ has well-known functions as a coenzyme in numerous redox reactions. The biochemistry of this molecule allows it to readily accept and donate electrons in reactions catalysed by enzymes of the mitochondrial electron transport chain, leading to the generation of ATP during oxidative phosphorylation (Rongvaux et al., 2003; Hassa et al., 2006). In addition to its function in energy metabolism, NAD⁺ serves as a substrate for covalent protein modifications by different ADP-ribosyl transferases, where the ADP-ribose moiety of NAD⁺ is enzymatically transferred onto acceptor proteins (Rongvaux et al., 2003). To avoid depletion of its intracellular pool, NAD⁺ is resynthesized by a *de novo* pathway and three distinct salvage pathways (reviewed by (Schreiber et al., 2006)). In brief, for *de novo* synthesis of NAD⁺, the essential aromatic amino acid L-tryptophan serves as a precursor being converted in several steps to nicotinic acid mononucleotide, then to nicotinic acid adenine dinucleotide and finally to NAD⁺. The three salvage pathways start either from nicotinic acid, nicotinamide riboside or nicotinamide. Nicotinic acid is converted through the so-called "Preiss-Handler pathway" to nicotinic acid mononucleotide which is connected with the *de novo* pathway (Preiss and Handler, 1958). Nicotinamide riboside and nicotinamide are both converted into the intermediate nicotinamide mononucleotide and finally to NAD^+ (Hassa et al., 2006; Schreiber et al., 2006).

1.4 Poly(ADP-ribose) metabolism

Poly(ADP-ribose) (PAR) is a heterogeneous linear or branched homopolymer of repeating ADP-ribose units coupled via ribose (1″→2′) ribose phosphate-phosphate bonds and displays several branching points resulting from the formation of ribose (1‴→2″) ribose linkages (Figure 2) (Bürkle, 2005; Hassa and Hottiger, 2008). Presently, seven members of the PARP superfamily have been shown to catalyse poly(ADP-ribosyl)ation i.e. PARP-1, PARP-2, PARP-3, vault-PARP, tankyrase-1, tankyrase-2 and PARP-7 (TiPARP) (Bürkle, 2005; Nguewa et al., 2005; Kleine et al., 2008). All these PARPs use NAD^+ as a substrate, the hydrolysis of which releases nicotinamide and covalently transfers the ADP-ribose moiety onto lysine or aspartic acid residues in the corresponding PARP enzyme (initiation). The transfer onto glutamic acid residues is up to date controversially discussed (Altmeyer et al., 2009; Tao et al., 2009). Then, the covalently bound ADP-ribose unit serves as a starting unit for linear and branched polymer elongation by adding further ADP-ribose moieties. The chain length can reach up to 200 units and branching generally occurs after every 20 to 50 units (Bürkle, 2005; Hassa and Hottiger, 2008). The molar ratio of the polymer is 1 adenine : 2 ribose : 2 phosphate, the product being susceptible to cleavage by phosphodiesterase but not by alkaline hydrolysis (David et al., 2009).

Poly(ADP-ribosyl)ation is a dynamic process as revealed by the short half-life of less than 1 min of the polymer in vivo (Whitacre et al., 1995). The transient nature of the polymer is largely due to its rapid degradation by a major catabolic enzyme poly(ADP-ribose) glycohydrolase (PARG; EC 3.2.1.143), discovered more than 30 years ago (Miwa et al., 1974) and existing in multiple isoforms with different subcellular locations. PARG cleaves ribose-ribose bonds in linear and branched regions of PAR to generate free ADP-ribose, whereas the ADP-ribosyl protein lyase or PARG itself remove protein-proximal single ADP-ribose units (Oka et al., 1984; Desnoyers et al., 1995; Davidovic et al., 2001). PARG can be processed into at least five mRNA splice variants in humans and mice, with resulting enzymes of hPARG-111/mPARG-110 (full-length 111/110 kDa protein in human (h) and mice (m)), hPARG-102 kDa (lack of exon 1), hPARG-99 kDa (lack of exon 1 and 2), hPARG-60/mPARG-63 (lack of exon 1-3 and 5/ lack of exon 1-3, respectively) and hPARG-55/mPARG-58 (lack of exon 1-3 and 5/ lack of exon 1-3, respectively), whereby only the full-length 111-kDa PARG is located in the nucleus due to a NLS in exon 1. The splice variants hPARG-102 and hPARG-99 are located in the cytoplasm, the hPARG-60/mPARG-63 splice variants in cytoplasm and nucleus, whereas hPARG-55/mPARG-58 are found in mitochondria

(Meyer-Ficca et al., 2004; Meyer et al., 2007). Furthermore, a further mammalian enzyme with poly(ADP-ribose) glycohydrolase activity, termed ADP-ribosyl hydrolase-3 (ARH3) but unrelated to PARG, has been identified (Oka et al., 2006). The degradation of PAR by PARG is essential for cell survival, as *PARG* gene disruption on exon 4 leads to massive PAR accumulation and embryo lethality at day (E) 3.5 (Koh et al., 2004). Recently, knockdown of PARG isoforms by stable and constitutive expression of a short hairpin RNA in HeLa cells resulted in beneficial effects in undamaged cells, as they were protected from spontaneous single-strand breaks and telomeric abnormalities (Amé et al., 2009).

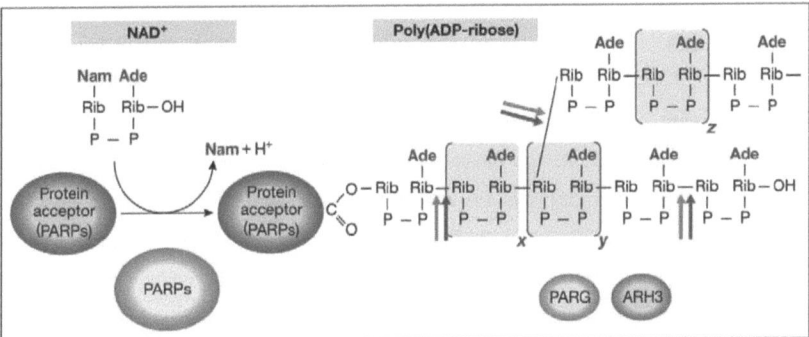

Figure 2: Schematic illustration of poly(ADP-ribose) metabolism. PARPs hydrolyse NAD^+ to ADP-ribose and nicotinamide and covalently attach the ADP-ribose moiety through an ester bond onto acceptor proteins or itself (initiation). PAR chains are generated by successive addition of ADP-ribose units and differ in size and complexity, as indicated by the x, y and z labels that represent values from 0 to 200 (elongation, branching). PAR is hydrolyzed by PARG and ARH3 at indicated positions. Ade, adenine; ARH3, ADP-ribosyl hydrolase-3; Nam, nicotinamide; PAR, poly(ADP-ribose); PARG, poly(ADP-ribose) glycohydrolase; Rib, ribose. Adapted from (Hakmé et al., 2008).

On the other hand, irradiation of these *PARG* deficient cells showed centrosome amplification leading to mitotic supernumerary spindle poles and accumulated aberrant mitotic figures, which caused either polyploidy or cell death (Amé et al., 2009).

1.5 Regulation of PARP activity

Several mechanisms have been shown to be involved in regulation of PARP activity, most of them acting on the post-translational level by modifications of PARP-1 itself. The regulation of PARP-1 principally can occur at different sites on the enzyme. PARP-1 is a highly abundant nuclear enzyme, which can be rapidly activated up to 500-fold by DNA strand breaks (D'Amours et al., 1999), whereas changes in its expression and/or abundance are regarded to have no primary

Introduction

regulatory relevance. Under certain pathophysiological conditions, however, up- and downregulation of PARP-1 protein levels have been reported, e.g. its upregulation in chronic heart failure (Pillai et al., 2005). Principally, PARP-1 is regulated at the level of its catalytic site. The most prominent and best characterized mechanism for a decrease in enzyme activity is mediated by lengthening of the polymer chain by auto-poly(ADP-ribosyl)ation of PARP-1 itself (Kawaichi et al., 1981). Moreover, the other product of PARP-1 action, nicotinamide, exerts a weak negative-feedback inhibition on its activity by binding to the catalytic center. In addition, several observations showed that phosphorylation of PARP-1 at specific serine/threonine residues also regulates its activity. Studies using microirradiation-induced DNA-damage have indicated, that altered phosphorylation at specific sites can modify the dynamics of assembly and disassembly of PARP-1 at sites of DNA damage (Gagne et al., 2009). Furthermore, it was demonstrated that phosphorylation of PARP-1, e.g. by extracellular signal-regulated kinases 1/2 (ERK1/2) or by c-Jun-N-terminal kinase 1 (JNK1), is required for a sustained and maximal PARP-1 activation after DNA damage (Kauppinen et al., 2006; Zhang et al., 2007). On the other hand, it was observed that PARP-1 loses DNA binding capacity and is inhibited by phosphorylation through protein kinase C (Tanaka et al., 1987; Bauer et al., 1992). PARP-1 can also be regulated by acetylation/deacetylation reactions, e.g. it has been shown that PARP-1 is activated through acetylation by p300/CREB-binding protein (Hassa et al., 2005). Furthermore, PARP-1 can be activated by mechanisms, which are unrelated to DNA damage and do not necessiate PARP-1 binding to DNA. Thus, Ca^{2+}-mediated activation of PARP-1 relates to the phospholipase C-inositol 1,4,5-triphosphate pathway (Homburg et al., 2000), moreover, PARP-1 can also be activated via a direct interaction with phosphorylated externally regulated kinase 2 (ERK2) (Cohen-Armon, 2007). Additionally, it was shown that the PARP-1 acetylation status is controlled by SIRT-1, a deacetylase and NAD^+-dependent enzyme promoting cell survival. As such, the polyphenolic stilbene derivative, resveratrol, proposed to exert a plethora of beneficial cardiovascular effects with antiproliferative properties (Stivala et al., 2001; Haider et al., 2003), has also been shown to stimulate SIRT-1, thus causing deacetylation of PARP-1 and consequently, inhibition of activity upon DNA damage and prevention of AIF-mediated cell death (Kolthur-Seetharam et al., 2006). Furthermore, SIRT-1 is able to protect stressed cardiomyocytes from PARP-1 mediated cell death by deacetylation of PARP-1 which is independent of DNA damage (Rajamohan et al., 2009). Recently, it was demonstrated that PARP-1 is sumoylated at the single lysine residue K486 within its automodification domain by small ubiquitin-like modifiers (SUMOs), SUMO1 and SUMO2, leading to inhibition of PARP-1 acetylation, which restrains the function of PARP-1 as a transcriptional coactivator of nuclear factor-kappaB (NF-κB) and hypoxia inducible factor 1 (HIF1) (Messner et al., 2009). Less is known regarding the regulation of basal PARP-1 activity, however, it is generally accepted that its high affinity binding to various DNA structures (e.g. cruciform, curved, nicked or supercoiled regions, three- and four-way junctions) might determine its activity status (Sastry and Kun, 1990; de Murcia and Menissier de Murcia, 1994; Lonskaya et al., 2005). Also permanently released oxidants and free radicals as by-products of oxidative phoshorylations and other pathways cause a low-level DNA strand breakage, thereby maintaining a basal PARP-1 activity (Wallace, 2001).

1.6 Role of PARP-1 in modulation of chromatin structure

In eukaryotic cells, DNA is packaged into chromatin, and this packaging impacts all DNA-dependent processes, including transcription (Nusinow et al., 2007). A growing body of evidence reveals that PARP-1 is an important key regulator of chromatin structure and transcription (Wacker et al., 2007). For instance, in the absence of the substrate NAD^+, PARP-1 binds to nucleosomes, causing chromatin condensation and transcriptional repression *in vitro*. *Vice versa*, in the presence of NAD^+, the automodification activity of PARP-1 is drastically stimulated by nucleosomes, which cause a release of PARP-1 from chromatin leading to a decompaction and thus restoration of transcription (Kim et al., 2004). It was demonstrated that both, the DBD and the catalytic domain of PARP-1, are required for efficient chromatin binding, compaction and transcriptional modulation. In detail, the DBD of PARP-1 was shown to be necessary for its binding to nucleosomes, whereby the catalytic domain cooperates with the DBD and promotes chromatin compaction and transcriptional repression independently of its enzymatic activity (Wacker et al., 2007). Moreover, studies have suggested that PARP-1 plays a role in regulating the composition of chromatin by modification of histones and other nuclear proteins, thereby influencing the regulation of gene expression (Kraus, 2008). Is was demonstrated that PARP-1 poly(ADP-ribosyl)ates and evicts the abundant chromatin-bound nuclear protein DEK from chromatin to permit access of the transcription machinery (Gamble and Fisher, 2007). Moreover, DEK is released from chromatin by extensive poly(ADP-ribosyl)ation in apoptotic cells in order to promote repair of DNA lesions and protect cells from genotoxic agents that typically trigger PARP-1 activation (Kappes et al., 2008). Furthermore, histone H1 and PARP-1 exhibit a reciprocal pattern of chromatin binding at many RNA polymerase II-transcribed promoters, leading to an enrichment of PARP-1 but depletion of histone H1 at these promoters. This characteristic differential pattern of PARP-1/H1 binding was associated with actively transcribed genes (Krishnakumar et al., 2008). Moreover, studies elucidating the interactions between PARP-1 and the non-histone domain of histone variant macroH2A (mH2A) have shown, that macroH2A1.2 (splice variant 1.2 from mH2A) promotes PARP-1 localization to chromatin and inhibits PARP-1 activity (Nusinow et al., 2007). Depletion of PARP-1 by RNA interference caused reactivation of a reporter gene on the inactive X chromosome, demonstrating that PARP-1 participates in the maintenance of silencing of the inactive X chromosome (Nusinow et al., 2007). Additionally, PARP-1 was identified as a part of the mH2A1.1 nucleosome complex (splice variant 1.1 from mH2A), which is associated with numerous gene promoters containing mH2A1.1 nucleosomes, particularly the promoter of heat shock protein HSP70.1. Upon heat shock the *HSP70.1* promoter-bound PARP-1 is released, thereafter transcription is activated through ADP-ribosylation of other *Hsp70.1* promoter-bound proteins (Ouararhni et al., 2006). Taken together these data are consistent with the histone shuttle model, in which PAR has a role in the transient removal of histones from DNA to facilitate DNA repair (Althaus, 1992).

1.7 DNA repair

DNA single-strand breaks (SSBs) can arise (i) by spontaneous hydrolytic degradation, (ii) from endogenous events including attack by reactive oxygen species (ROS) released by cellular metabolic processes, or (iii) by alkylating agents and ionizing radiation (IR) leading to damage of sugar residues and disintegration of the DNA backbone. As SSBs can lead to potentially harmful mutations and genomic instability, the cell has evolved a variety of strategies to repair DNA damage. One of them is the repair of SSBs by base excision repair (BER), which acts on a wide variety of DNA lesions. In each mammalian cell, more than tens of thousands of SSBs arise each day, both directly from disintegration of damaged sugars, and indirectly from the base excision repair (BER) of damaged bases (Caldecott, 2001). SSBs generally have blocked or damaged termini that lack the conventional 5´-phosphate and the 3´-hydroxyl groups (Horton et al., 2008). Unrepaired SSBs can disrupt transcription and replication or can be converted into lethal DNA double-strand breaks (DSBs) at stalled replication forks, which can lead to chromosome fragmentation and cell death (Caldecott, 2004).

1.7.1 Base excision repair (BER) and involvement of PARP-1/PARP-2

One of the best studied roles of PARP-1 and PARP-2 is their involvement in base excision repair (BER). BER is activated by single DNA base mutations, caused by oxidation, deamination and alkylation. Principally, BER can be divided into a so-called short-patch BER pathway where only a single nucleotide is replaced, and a long-patch BER pathway where 2 to 13 nucleotides are incorporated (Hakem, 2008). The involvement of PARP-1 and PARP-2 in the BER pathway was shown by treatment of *Parp-1$^{-/-}$* or *Parp-2$^{-/-}$* mouse embryonic fibroblasts (MEFs) with alkylating agents, which demonstrated severe defects in DNA strand break repair (Trucco et al., 1998; Beneke *et al.*, 2000a; Dantzer *et al.*, 2000; Masutani *et al.*, 2000; Schreiber *et al.*, 2002). *Parp-1*$^{-/-}$ cells showed a delayed strand break resealing after methyl methanesulfonate (MMS) treatment, and *Parp-1*$^{-/-}$/DNA polymerase beta$^{-/-}$ (polβ) double mutant cell lines were greatly affected in repairing 8-oxo-7,8-dihydroguanine (8-oxoG) damage completely (Trucco *et al.*, 1998; Le Page *et al.*, 2003). In response to suffered SSBs, a rapid auto-poly(ADP-ribosyl)ation of PARP-1 is necessary for recruitment of the molecular scaffold protein X-ray repair complementing factor 1 (XRCC1) to the SSB. XRCCI contains a BRCT1 domain which interacts with PARP-1 and PARP-2, additionally being supplied with a binding site for PAR, which is indispensable for its poly(ADP-ribose)-dependent recruitment to the SSBs (El-Khamisy *et al.*, 2003; Okano *et al.*, 2003). In addition, it was shown that PARP-1 is indispensably involved in the long-patch BER pathway, as the repair of abasic sites was about half as efficient in PARP-1-deficient cell extracts compared to wild type cell extracts, particularly at the polymerization step of the short-patch repair synthesis, but were nearly inefficient at the long-patch repair (Dantzer *et al.*, 2000). Moreover, it was shown that PARP-1 along with flap endonuclease-1 (FEN-1) stimulates strand displacement DNA synthesis by DNA Polβ in the long-patch BER pathway (Prasad *et al.*, 2001). Recently, it was demonstrated that PARP-1 interacts with the DNA repair protein aprataxin and is required for its recruitment to sites of DNA strand breaks (Harris *et al.*, 2009). Furthermore, it was shown that both, PARP-1 and PARP-2 can homo-and heterodimerize and interact both with multiple nuclear components of the SSB repair and BER machinery, including XRCC-1, DNA polβ and DNA ligase III (Schreiber *et al.*, 2002). However, it appears that PARP-2 has a different role and acts with different kinetics in the BER pathway, as its recruitment to DNA damage sites succeeds that of PARP-1. Other studies show that PARP-2 does not recognize SSBs but rather gaps or flap structures, indicating that PARP-2 may have its function in later steps of the DNA repair process (Mortusewicz *et al.*, 2007).

1.7.2 Nucleotide excision repair (NER)

Nucleotide excision repair (NER) is the pathway that removes ultraviolet (UV) radiation-induced photoproducts such as pyrimidine dimers or 6-4 photoproducts, otherwise causing helical distortion and bending from the DNA (Tremblay *et al.*, 2009). In response to UV radiation, mammalian cells rapidly activate PARP-1, and it was shown that one of the causes for its activation is DNA damage, such as formation of thymine dimers, which are repaired by the NER process (Vodenicharov *et al.*, 2005). This finding was confirmed by PARP-1 RNA interference experiments in human skin fibroblasts, where UV radiation reduced host cell reactivation of a UV-damaged adenovirus-encoded reporter gene (Ghodgaonkar *et al.*, 2008). Furthermore, one of the core factors in the NER incision complex, namely xeroderma pigmentosum group A (XPA) protein, was identified as

Introduction

poly(ADP-ribose) binding protein (Fahrer *et al.*, 2007). Recently, it was demonstrated that PARP-1 has strong affinity for platinum-damaged DNA, e.g. evoked by the anticancer drug cisplatin, which is repaired by NER, the major mechanism for removing cisplatin adducts from DNA (Guggenheim *et al.*, 2009).

1.7.3 Double-strand break (DSB) repair

DNA double-strand breaks mostly arise from ionizing radiation (IR) (1 Gy induces approx. 40 DSBs and 1000 SSBs per cell (Pandita and Richardson, 2009)), ROS, chemicals, as intermediates during V(D)J recombination (Bassing *et al.*, 2002) and immunoglobulin class-switch recombination (Honjo *et al.*, 2002), or occur during replication when replication forks encounter DNA SSBs. DSBs are the most harmful form of DNA damage and if unrepaired, can evoke chromosomal aberrations, as a single DSB bears the risk to induce cell death (Rich *et al.*, 2000; Pfeiffer *et al.*, 2004). As a consequence, the mammalian cell has evolved two main mechanisms for repair of DSBs: (i) non-homologous end-joining (NHEJ) and (ii) homologous recombination (HR). The major repair pathway is the non-homologous end-joining (NHEJ), characterized by two broken DNA ends directly rejoined, regardless of the phase of cell cycle and independent of homologous sequences (Yano *et al.*, 2009). HR promotes an error-free repair and is active only in the S/G2 phase of the cell, since it requires the sister chromatid necessary to function as a template for DNA repair (Pardo *et al.*, 2009). *In vitro* experiments revealed binding sites for PAR in two critical NHEJ pathway proteins, namely (i) the DNA-dependent protein kinase catalytic subunit (DNA-PKcs), a heterotrimeric enzyme required for rejoining of DSBs, and (ii) Ku70, an essential component of a protein complex that also binds to DSBs and activates DNA-PK (Pleschke *et al.*, 2000). Further functional interactions between PARP-1 and DNA-PK have been identified by *in vitro* studies, demonstrating that DNA-PK is capable to phosphorylate PARP-1, and in turn, PARP-1 poly(ADP-ribosyl)ates the DNA-PKcs and stimulates its activity (Ruscetti *et al.*, 1998). Recently, it was demonstrated that PARP-1 together with DNA ligase III operates in an alternative, more error-prone backup pathway of NHEJ (Wang *et al.*, 2006). Furthermore, PARP-1 binds to and is activated at stalled replication forks, attracting Mre11, a member of the MRN complex (MRE11/Rad50/NBS1), which is the beginning step in HR and required for end-processing, and then promotes homologous recombination and replication restart (Bryant *et al.*, 2009). However, despite its involvement in NHEJ pathway proteins, PARP-1 appears to have rather a regulatory function for decision between the two DSB repair pathways. Consistent with this, attachment of PAR to Ku70/80 (involved in NHEJ) reduces its affinity to DSBs *in vitro* (Li *et al.*, 2004). Furthermore, PARP-1 and the

Introduction

postreplicative repair protein Rad18 are independently capable to facilitate HR and suppress NHEJ at stalled replication forks (Saberi et al., 2007).

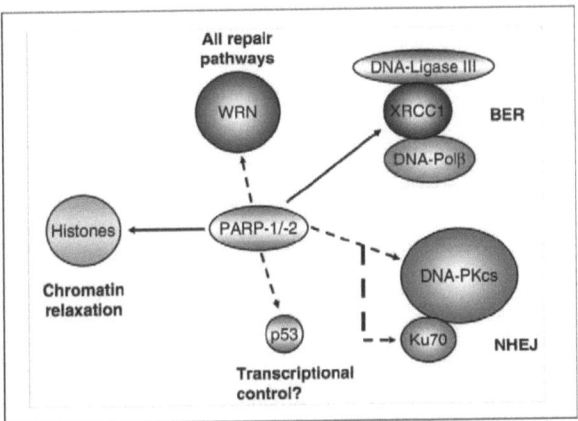

Figure 3: The role of PARP-1/-2 in DNA repair. PARP-1/-2 and/or PAR interact with several proteins involved in BER and NHEJ. Histones are poly(ADP-ribosyl)ated by PARP-1/-2 leading to chromatin relaxation. PARP-1 is able to modulate the activity of the Werner syndrome protein (WRN) and inhibits transcriptional activity of p53 by poly(ADP-ribosyl)ation. A full-line arrow indicates interactions shown for both PARPs, whereas dashed arrows indicate established interactions between PARP-1 and other proteins. Adapted from (Beneke and Bürkle, 2007).

1.8 Implication of PARP in aging, longevity and genomic stability

Accumulation of DNA damage, genomic instability and macromolecular damage are suggested to act as a driving force for the aging process (Csiszar et al., 2005). The implication of PARP in aging was demonstrated by measurement of maximal PARP activity in permeabilised mononuclear leukocytes from different mammalian species following stimulation with saturating amounts of double-stranded oligonucleotide and NAD^+ (Grube and Bürkle, 1992). As a result, they found a strong positive correlation between maximum PARP activity and maximum lifespan of 13 different mammalian species, e.g. a 5-fold higher maximal PARP activity in humans compared to rats, irrespective from the PARP protein level present, indicating a greater specific PARP activity in longer-lived species and being proposed to rely on slight variations in primary structure of the enzyme of each species investigated. These findings fit well with the positive correlation between the rate and extent of unscheduled DNA synthesis (as a measure of excision-repair) after UV radiation of seven different mammalian fibroblasts, which increase with the species-characteristic lifespan (Hart and Setlow, 1974). Moreover, intraspecies comparisons of maximal PARP activity as a function of chronological age in humans and rats revealed a decline with age (Grube and Bürkle,

1992). The decline of PARP activity during the aging process was also supported by the finding of a reduced PARP activity in nuclear fractions from hippocampus of old rats compared to young rats (Strosznajder *et al.*, 2000). Consistent with the implication of PARP-1 in aging, *Parp-1$^{-/-}$* mice showed accelerated aging and a reduction of lifespan (Piskunova *et al.*, 2008). The primary structure of PARP-1 is highly conserved in eukaryotes, e.g. between human and mouse exists a homology of 92% at the level of amino acid sequence (Virág and Szabó, 2002). A recombinant expressed and purified human PARP-1 protein showed a two-fold higher automodification activity compared to the respective rat protein, thus emphasizing a structure-dependent difference in enzyme activity from both species (Beneke *et al.*, 2000b). Within the *hPARP-1* gene several single nucleotide polymorphisms are known, one of them comprises exchange of valine (aa762) by alanine in the catalytic domain, causing a reduction in poly(ADP-ribosyl)ation activity of nearly 50%, associated with an increased risk of lung cancer for people affected with this polymorphism (Cottet *et al.*, 2000; Zhang *et al.*, 2005; Wang *et al.*, 2007). However, so far the Val762Ala polymorphism was not shown to be associated with the age of centenarians (Cottet *et al.*, 2000). Compared with the results obtained in mononuclear blood cells, these findings could not fully account for the difference in PARP-1 activity. Therefore, along with several single nucleotide polymorphisms further impacts that modulate PARP-1 activity, like species specific posttranslational modifications or different accessory proteins, might additionally be considered (Beneke and Bürkle, 2004).

Interestingly, high poly(ADP-ribosyl)ation activity is positively correlated with human longevity, demonstrated by measurement of oligonucleotide-stimulated maximal PARP activity in permeabilised lymphoblastoid cells, which revealed a significant higher maximal PARP activity in centenarians compared to 20-70 years old controls (Muiras *et al.*, 1998). Further evidence for a link between aging/longevity and PARP-1 was recently presented and confirmed by experiments, in which PARP activity in cells from human (HeLaS3, IMR90) and hamster (COM3), either inhibited by 3-aminobenzamide or specifically knocked down by RNA interference, resulted in a rapid decrease in median telomere length, which could be restored to control levels after the removal of inhibitor (Beneke *et al.*, 2008).

Furthermore, it is hypothesized that PARP-1 acts as a negative regulator of genomic instability (Meyer *et al.*, 2000; Bürkle, 2001). This was strengthened by the observation that conditional overexpression of hPARP-1 in stably transfected hamster cells (COMF10) caused several-fold stimulation of PAR formation after γ-irradiation, suppressed the rate of sister chromatid exchanges after N-methyl-N´-nitro-N-nitrosoguanidine (MNNG) treatment (Meyer *et al.*, 2000), accompanied by a decrease in micronucleus formation after treatment with MNNG (Diploma thesis, Raphael Hahn, 2004, University of Konstanz), MMS or bleomycin (Diploma thesis, Yvonne Rüdigier, 2005,

University of Konstanz). *Vice versa, trans*-dominant overexpression of the 42-kDa DBD of hPARP-1 caused a reduction of PAR formation after γ-irradiation by approximately 90% (Kupper *et al.*, 1995) and increased genomic instability under genotoxic stress, examined after exposure of the cells to MNNG and characterized by increase in gene amplification (Kupper *et al.*, 1996). Furthermore, DBD-overexpression raised the rates of spontaneous as well as MNNG-induced sister chromatid exchanges (Schreiber *et al.*, 1995), and increased the number of micronuclei after treatment with bleomycin (Diploma thesis, Raphael Hahn, 2004, University of Konstanz), demonstrating an important role of PARP-1 in genomic stability.

Finally, PARP-1 was shown to interact and modulate the activity of the RecQ DNA helicase Werner syndrome protein (WRN), a genetic premature aging disorder in which the *WRN* gene is mutated (Adelfalk *et al.*, 2003; Lebel *et al.*, 2003; von Kobbe *et al.*, 2004). Cells from patients suffering from Werner syndrome are affected by genomic instability, defects in replication, altered telomere dynamics and additionally, many WRN-interacting proteins are involved in BER and NHEJ suggesting an important function in the DNA repair process (Opresko *et al.*, 2003; von Kobbe *et al.*, 2004). Primary cells obtained from Werner syndrome patients and treated with H_2O_2 or MMS, revealed a severely defective poly(ADP-ribosyl)ation of nuclear proteins other than PARP-1, indicating that a functional WRN/PARP-1 complex is required for poly(ADP-ribosyl)ation of nuclear proteins, which plays a key role in the cellular response to oxidative stress and alkylating agents (von Kobbe *et al.*, 2003).

1.9 PARP and its implication in T-cell development

The discovery that PARP activity is higher in B-cell lines from centenarians than in younger controls, has led to the hypothesis that PARP-1 contributes to genomic maintenance of lymphocytes and as a consequence, may influence immunosenescence (Muiras *et al.*, 1998). Furthermore, it was found that *Parp-2*$^{-/-}$ but not *Parp-1*$^{-/-}$ mice, had a reduction in $CD4^+CD8^+$ thymocyte cell number by a factor of 2, which was associated with several abnormalities, such as a decreased $CD4^+CD8^+$ cell survival, a skewed repertoire of T-cell receptor α toward the 5′Jα segments, and an increased expression of proapoptotic factors (Yelamos *et al.*, 2006). These authors proposed a model, in which the absence of PARP-2 affects the repair of DSBs generated during Vα to Jα rearrangements, which in turn could activate apoptosis, leading to a reduced lifespan and an impairment in the formation of secondary Vα to Jα rearrangements. Moreover, enzymatic activity of PARP-1 seems to be necessary for transcription of several T helper cell 1 (Th-1) cytokines, such as interleukin-2 (Il-2), interferone-γ (INF-γ) and tumor necrosis factor-alpha (TNFα), which are necessary for lymphocyte activation (Chiarugi, 2002; Chiarugi and Moskowitz, 2003; Maruyama *et al.*, 2007).

Introduction

The activation of T lymphocytes, which results in their proliferation and differentiation, requires a co-stimulatory signal of the antigen-specific T-cell receptor (TCR)/CD3 complex and CD28 receptor, which leads to a ligation of these receptors. The ligation then initiates signal cascades and activates downstream transcription factors, such as NF-κB, activator protein 1 (AP-1) and nuclear factor of activated T-cells (NFAT), which play a critical role in reprogramming gene expression (Saenz et al., 2008). Recently, it has been demonstrated that PARP-1 is activated during T-cell stimulation independently of DNA damage, binds to and then ADP-ribosylates NFAT, a transcription factor family, which is pivotal for T lymphocyte functionality and interleukin-2 (IL-2) expression (Valdor et al., 2008). The ADP-ribosylation of NFAT was shown to increase NFAT binding to DNA, thereby enhancing NFAT-mediated IL-2 expression, as pharmacological inhibition or genetic ablation of PARP-1 reduced NFAT-dependent IL-2 and IL-4 cytokine expression in T-cells (Olabisi et al., 2008). These data suggest that PARP-1 is capable to modulate immune functions by acting as a positive co-regulator of NFAT-dependent cytokine gene expression in T-cells. In order to get deeper insights into the transcriptional regulation by PARP-1 during reprogramming of gene expression that takes place upon activation of T-cells, gene expression studies had been performed in stimulated splenic T-cells derived from $Parp-1^{+/+}$ or $Parp-1^{-/-}$ mice (Saenz et al., 2008). As a result, these experiments revealed a significant increase in expression of IL-4 in $Parp-1^{+/+}$ cells and besides other cytokines, a reduced expression level of IL-4 and INF-γ in $Parp-1^{-/-}$ cells. These data suggest an important role for PARP-1 to bias T-cell response to a Th-2 phenotype, as IL-4 is a major Th-2 effector cytokine and a key promoter of Th-2 development, whereas INF-γ represents the major effector cytokine of Th-1 cells (Saenz et al., 2008). Finally, it was demonstrated that T-cell proliferation, which is a hallmark of activated T-cells, is significantly impaired in the absence of PARP-1, accompanied by upregulation of a lot of genes responsible for lymphocyte activation in $Parp-1^{+/+}$ cells compared to $Parp-1^{-/-}$ cells (Saenz et al., 2008).

1.10 The role of PARP-1 in cell survival and cell death

PARP-1 can act either as an indispensible factor for cell survival or, following its overactivation, as a mediator of cell death. Particularly, in response to mild to moderate genotoxic stimuli, PARP-1 activation facilitates DNA repair and cell survival without the risk of remaining mutated genes. However, after severe and excessive stimuli that cause a corresponding extent of DNA damage, cell death can occur via apoptosis and necrosis, respectively. The three pathways determining the fate of cells related to the intensity of DNA-damaging stimuli, and the quantitatively different involvement of PARP-1 is depicted in Figure 4. Generally, necrosis is induced by unphysiological excessive DNA-damaging stimuli, e.g. caused by oxidative or nitrosative stress to cells, whereas apoptosis is induced by more physiological mild stimuli that eliminates cells with severe and not repairable DNA damage in order to maintain normal tissue homeostasis. Apoptosis is an active, energy- (ATP) dependent process of programmed cell death composed of a number of characteristic steps, involving compaction of affected cells and their elimination by macrophages (reviewed in (Bratton and Cohen, 2001)), whereas necrosis is a detrimental passive process, characterized by disintegration of the plasma membrane, followed by leakage of cell content into the surrounding tissue, thereby exacerbating inflammation.

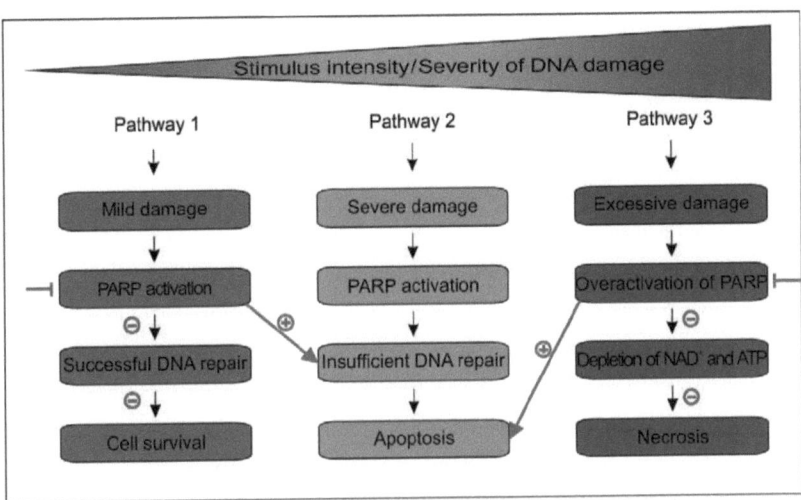

Figure 4: The role of PARP activity in cell survival and cell death in response to DNA damage intensity. After low levels of DNA damage, PARP facilitates DNA repair and acts as a cell survival factor (pathway 1). In contrast, after severe, unrepairable or excessive DNA damage, PARP promotes either apoptosis (pathway 2) or necrosis by PARP overactivation and depletion of NAD^+ and ATP (pathway 3). Pharmacological inhibition of PARP (indicated in blue) in cells entering pathway 1 inhibits DNA repair, prevents cell survival, directs cells into pathway 2 and results in apoptosis, whereas PARP inhibition in cells entering pathway 3 preserves cellular energy, leading to a shift from necrosis to apoptosis. Adapted and modified from (Jagtap and Szabó, 2005).

In addition to the intensity of DNA-damaging stimuli which determines the fate of cells, *i.e.* their survival or death (Figure 4), the metabolic status of the cell is considered a further important factor that modulates the choice of PARP-1-dependent pathways by which cell death, either via apoptosis or necrosis, can be mediated (Nosseri *et al.*, 1994; D'Amours *et al.*, 2001; Kim *et al.*, 2005). In actively proliferating cells including cancer tissues, cellular levels of ATP synthesized in the cytoplasm by the glycolytic pathway are diminished in response to excessive PARP-1 activation and depletion of cellular NAD^+ levels, thereby preventing the apoptotic pathway from functioning, and finally leading to cell death by necrosis, a process which does not require ATP to proceed. In contrast, in non-proliferating cells, the mitochondrial oxidative phosphorylation still preserves the cellular ATP supply, enabling these cells to undergo the more ordered cell death via apoptosis (Zong et al., 2004).

Overactivation of PARP-1 has been convincingly shown to be implicated in cell death in a variety of pathophysiological conditions, including tissue damage by ischemia and reperfusion (Eliasson et al., 1997; Endres et al., 1997; Pieper et al., 2000; Hendryk et al., 2008; Kauppinen et al., 2009; Sodhi et al., 2009), excitotoxicity (Lipton and Rosenberg, 1994; Mandir et al., 2000) and various inflammatory processes (Gonzalez-Rey et al., 2007; Kim et al., 2008; Naura et al., 2008; Jog et al., 2009).

As PARP-1 is an NAD^+ consuming enzyme, it is hypothesized that it may act as a molecular switch diverting cells to undergo either apoptosis or necrosis. During the execution phase of apoptosis, PARP-1 is cleaved at the DEVD sequence within the NLS region by caspase-3 and caspase-7, to produce a 24-kDa fragment, which retains the DNA binding domain, and an 85-kDA fragment with reduced catalytic activity (Lazebnik et al., 1994; Tewari et al., 1995; Germain et al., 1999). Both fragments are able to inhibit PARP activity, as they inhibit homodimerization of PARP-1 as well as binding of intact PARP-1 to DNA (Kim et al., 2000a; Kim et al., 2000b; D'Amours et al., 2001). Overexpression of the apoptotic DNA-binding domain of PARP-1 provided evidence that it inhibits the catalytic activity of uncleaved PARP-1 in a dominant-negative manner (Kupper et al., 1990; Molinete et al., 1993; Kupper et al., 1995; Schreiber et al., 1995). This caspase-mediated inactivation of PARP-1 is suggested to maintain cellular energy required for certain ATP-sensitive steps in the execution phase of apoptosis. *Vice versa*, cells expressing a caspase-noncleavable PARP-1 mutant showed an accelerated necrotic cell death after treatment with TNFα (Herceg and Wang, 1999; Los et al., 2002). Therefore, it was postulated that one reason for PARP-1 cleavage during apoptosis, is to prevent survival of severely damaged cells (Halappanavar et al., 1999). Further evidence for an involvement of poly(ADP-ribosyl)ation reactions in necrotic cell death, due to an overstimulation PARP-1, has been provided by several studies, including high-dosage alkylating agent treatment with MNNG (Pogrebniak et al., 2003; Liu et al., 2008), H_2O_2 (Watson et al., 1995; Filipovic et al., 1999), or peroxidizing peroxynitrite (reviewed in (Korkmaz et al., 2007)), revealing that pharmacological inhibition of PARP activity or knock-out of the *PARP-1* gene prevent cell death through the necrotic route. Moreover, it was demonstrated that overactivation of poly(ADP-ribosyl)ation reactions mediates cell death in a caspase-independent pathway, whereby PAR triggers the release of apoptosis-inducing factor (AIF) from mitochondria into the cytoplasm and its translocation to the nucleus, followed by chromatin condensation, DNA fragmentation, nuclear shrinkage and finally, cell death (Andrabi et al., 2006; Yu et al., 2006). The release of AIF and its translocation was blocked in cells, whose PARP activity was pharmacologically inhibited, or in cells lacking the PARP-1 protein (Yu et al., 2002; Chen et al., 2004; Xiao et al., 2005).

1.11 PARP-2

More than 30 years after the discovery of PARP-1, a second DNA damage-dependent isoform, PARP-2, was found based on the residual presence of 5% to 10% of the DNA-dependent PARP activity in $Parp-1^{-/-}$ cells (Shieh et al., 1998; Amè et al., 1999; Schreiber et al., 2002). The *PARP-2* gene encodes a 62-kDa protein that shares considerable homology with *PARP-1*, which in the catalytic domain accounts for 60%. Compared to PARP-1, PARP-2 contains a shortened DNA-binding domain composed of only 64 amino acids lacking any DNA binding motif, but displays automodification properties similar to PARP-1. Differences in the structure of PAPR-1 and PARP-2 mostly reside in the proximity of the PAR acceptor site, suggesting that both isoforms may specifically interact with different substrate proteins (Oliver et al., 2004). Both PARP-1 and PARP-2 are localized in the nucleus, become activated by DNA strand breaks and are able to homo- and heterodimerize (Huber et al., 2004; Bürkle, 2005). PARP-2 is suggested to play an important role in BER, as it was found to interact with three other proteins involved in the BER pathway: XRCC1, DNA polβ and DNA ligase III, already known as partners of PARP-1 (Schreiber et al., 2002). In addition, PARP-2 deficient mouse embryonic fibroblasts treated by the alkylating agent N-nitroso-N-methylurea (MNU) or subjected to ionizing radiation, displayed a significant delay in DNA strand break resealing, similar to that observed in PARP-1 deficient cells (Schreiber et al., 2002; Menissier de Murcia et al., 2003). Currently, available evidence emerging from numerous studies ascribes functions of PARP-2 which are similar to those of PARP-1, as PARP-2 can compensate in many ways for a deficiency of PARP-1 (Chalmers et al., 2004). Although both enzymes share many complementary functions, recently however, important functional differences have been detected. Particularly, a selective inhibition of PARP-2 with UPF-1069 (an isoquinolinone derivative) enhanced oxygen-glucose deprivation (OGD) induced cell death in rat hippocampal slices as a model for a caspase-dependent, apoptosis-like process, whereas inhibition of both PARP-1 and PARP-2 with TIQ-A (thieno[2,3-c]isoquinolin-5-one) had no effect on neuronal survival (Moroni et al., 2009). Thus, it seems that PARP-2 activation helps hippocampal pyramidal cells to survive, whereas its selective inhibition exacerbates cell damage in this brain area similarly as previously observed in $Parp-2^{-/-}$ mice exposed to short periods of cardiac arrest (Kofler et al., 2006). In contrast, selective inhibition of PARP-2, or unselective inhibition of both PARP-1 and PARP-2, increased cell survival after OGD in murine mixed cortical cell cultures, an *in vitro* stroke model characterized by a necrosis-like process (Moroni et al., 2001; Moroni et al., 2009). The observation that selective PARP-2 inhibition appeared to be sufficient to reduce the extent of OGD-evoked cell death in mouse cortex cells *in vitro*, is in agreement with previous findings *in vivo* showing that *Parp-2$^{-/-}$* mice had a reduced brain infarct after middle cerebral occlusion (Kofler et al., 2006). *Parp-2* knock-out mice are viable (Menissier de Murcia et al., 2003) as their *Parp-1* counterparts (Wang et al., 1995). Similar to PARP-1, also PARP-2 is susceptible to cleavage by caspases (mainly caspase-8) during apoptosis (Benchoua et al., 2002), and cells lacking both isoforms are characterized by higher incidence of DNA fragmentation and accelerated apoptosis than wild type cells (Heyer et al., 2000). According to the present view, it is hypothesized that neither PARP-1 nor PARP-2 are necessary for normal growth and development, at least regarding mouse, although the lack of either protein leads to genomic instability and higher sensitivity to alkylating agents or irradiation. Taken together, these studies indicate that PARP-1 and PARP-2 possess both overlapping and non-redundant functions for normal growth and development as well as preserving genomic stability.

1.12 PARP-1 and PARP-2 mediated functions in knock-out mice

To elucidate the biological function of poly(ADP-ribosyl)ation and PARP-1/PARP-2 *in vivo*, three independent *Parp-1* knock-out mice (Wang et al., 1995; de Murcia et al., 1997; Masutani et al., 1999) and one *Parp-2* knock-out mouse (Menissier de Murcia et al., 2003) were generated. *Parp-1$^{-/-}$* mice from all three groups displayed a normal phenotype, were viable, of normal growth and fertile, indicating that PARP-1 is not critical for development. Only one group observed that older mice developed obesity and epidermal hyperplasia by 30% compared to wild type (Wang et al., 1995). However, in experiments with 4-week-old *Parp-1$^{-/-}$* mice, intraperitoneal injection of the alkylating agent MNU, caused death of all 19 animals within 4 weeks, whereas only 9 from 24 *Parp-1$^{+/+}$* mice died in the same time period (de Murcia et al., 1997). In addition, these authors also performed whole body γ-irradiation studies (8 Gy), in which *Parp-1$^{-/-}$* mice rapidly died due to acute radiation toxicity observed in the small intestine, whereas all *Parp-1$^{+/+}$* mice remained unaffected, showing that PARP-1 acts as a obligatory survival factor (de Murcia et al., 1997). *Parp-1$^{-/-}$* mice investigated by the Sugimura group also demonstrated enhanced sensitivity to treatment with MNU or γ-irradiation compared to wild type mice, assigning PARP-1 a protecting role in response to high levels of DNA damage (Masutani et al., 2000).

Studies with proliferating *Parp-1$^{-/-}$* cells exposed to γ-irradiation revealed an impaired growth (Wang et al, 1995). *Parp-1* knock-out disclosed an important role of PARP-1 for maintenance of genomic integrity, as primary *Parp-1$^{-/-}$* mouse mutant fibroblasts and splenocytes treated with mitomycin C or γ-irradiation, exhibited an elevated rate of spontaneous sister chromatid exchanges and an increased formation of micronuclei (Wang *et al.*, 1997). Micronuclei formation is considered as a marker of chromosome instability closely related to double-strand break induction, reflecting the ability of the cell to sequester DNA lesions in the micronuclear bodies. DNA repair studies using the alkaline comet assay demonstrated a slower DNA repair rate in $Parp-1^{-/-}$ MEFs after treatment with MMS compared to wild type cells, and in addition, showed growth retardation, G2/M accumulation and chromosome instability, which strengthened the implication of PARP-1 in the base-excision repair (BER) process (Trucco *et al.*, 1998). Taken together, these data support the obligatory presence of PARP-1 in cells exposed to genotoxic stress above endogenous levels.

$Parp-2^{-/-}$ mice did not display any abnormal phenotype, however, they showed a higher sensitivity to whole body γ-irradiation treatment compared to wild type mice, but to a lesser extent than $Parp-1^{-/-}$ mice (Menissier de Murcia *et al.*, 2003). Treatment of bone marrow cells taken from $Parp-2^{-/-}$ mice with MNU, resulted in increased post-replicative genomic instability accompanied by elevated sister chromatid exchanges, G2/M accumulation, chromosome mis-segregation and kinetochore defects. Irradiation of these cells with 2 Gy increased the number of chromatid breaks, preferentially in centromeric regions, compared to wild type cells (Menissier de Murcia *et al.*, 2003). These findings suggest that both PARP-1 and PARP-2 enzymes play an important role in maintaining genomic stability under genotoxic stress.

In order to get further insights into the functional interaction of PARP-1 and PARP-2, efforts to generate double mutant $Parp-1^{-/-}/Parp-2^{-/-}$ mice were undertaken but remained unsuccessful, as these animals were not viable and died early at the onset of gastrulation, *i.e.* at a development stage the embryo normally undergoes rapid cellular proliferation, and at which it is highly sensitive to DNA damage and prone to death (Heyer *et al.*, 2000; Menissier de Murcia *et al.*, 2003). Taken together, these results disclose that both isoforms, PARP-1 and PARP-2, possess overlapping as well as non-redundant functions in the maintenance of genomic stability and that both enzymes are obligatory for normal growth and development.

1.13 Pharmacological inhibition of PARP

Besides the DNA-repairing and cytoprotective effect of PARP activity, which is apparent under conditions of low to moderate DNA damage, a more intensive stimulation of PARP in response to abundant genotoxic stimuli activates an apoptotic pathway to eliminate cells with insufficiently repaired DNA, mediated via release of AIF from mitochondria (Yu *et al.*, 2002). However, severe DNA damage or consequences of a variety of cardiovascular and inflammatory diseases, such as shock, ischemia, diabetes, and neurodegenerative disorders, can cause excessive activation of

Introduction

PARP, which depletes the intracellular pools of NAD^+ and subsequently ATP, ultimately leading to cellular dysfunction and necrosis by rapid energy consumption (Pieper et al., 1999b; Virág and Szabó, 2002; Amè et al., 2004). Consequently, depending on the circumstances, pharmacological inhibitors of PARP have the potential to either enhance the cytotoxicity of antitumor treatment by depressing DNA repair, and thus leading to apoptosis of tumor cells, or to provide remarkable protection from tissue damage in various forms of reperfusion organ injury, inflammation, and neurotoxicity in animal models (Virág and Szabó, 2002; Beneke and Bürkle, 2004; Jagtap and Szabó, 2005; de la Lastra et al., 2007). The observation that nicotinamide itself, a product of PARP catalytic activity, is a weak PARP inhibitor, originated to the development of PARP inhibitors (Lord and Ashworth, 2008). A further competitive inhibitor is 3-aminobenzamide (3-AB) (Figure 5), which blocks NAD^+ binding to the catalytic domain of PARP, but besides its low potency and limited cellular uptake, 3-AB was shown to be a strong scavenger of free radicals (Wilson et al., 1984; Virág and Szabó, 2002). Meanwhile, more potent and selective PARP inhibitors have been discovered, most of them bearing the classic benzamide structure mimicking to some degree the nicotinamide moiety of the substrate NAD^+, but none of them were able to distinguish between the two isoforms PARP-1 and PARP-2 by a factor of ten or more. N-(6-Oxo-5,6-dihydro-phenanthridin-2-yl)-N,N-dimethylacetamide (PJ34) (Figure 5) is an example of a stronger PARP inhibitor, exceeding potency of 3-AB by a factor of approximately 1,000, additionally being supplied with good water-solubility but no selectivity for either PARP-1 or PARP-2. However, since distinct binding modes that are necessary for discrimination between ligands and each isoenzyme have been discovered, the synthesis of PARP-1 selective quinazolinones and PARP-2 selective quinoxalines became possible (Iwashita et al., 2004a; Iwashita et al., 2004b; Ishida et al., 2006). Most recently, the synthesis of a 5-benzoyloxyisoquinolin-1(2H)-one derivative, UPF-1069, as the most selective PARP-2 inhibitor known so far (selectivity index >60) was reported (Pellicciari et al., 2008; Moroni et al., 2009).

PARP-1 activation contributes to the tissue injury caused by ischemia and reperfusion in various organs, including heart (Eliasson et al., 1997; Thiemermann et al., 1997; Liaudet et al., 2001). A reduction in infarct size and/or improved cardiac contractility after myocardial ischemia in rats has been demonstrated for PARP inhibitors of different chemical structure, e.g. nicotinamide (NA), 3-aminobenzamide (3-AB), 4-hydroxyquinazoline (4-HQN), 1,5-dihydroxyisoquinoline (ISQ), 5-aminoisoquinolin-1(2H)-one (5-AIQ), 1,11b-dihydro-[2H]benzopyrano[4,3,2-de]isoquinolin-3-one (GPI-6150), N-(6-oxo-5,6-dihydrophenanthri-din-2-yl)-N,N-dimethylacetamide (PJ34) and (6-fluoro-2,3,4,11b-tetrahydro-1H-fluoreno[1,9-cd]azepin-10-ylmethyl)-methyl-amine (INO-1001) (Thiemermann et al., 1997; Zingarelli et al., 1997; Bowes et al., 1998; Docherty et al., 1999; McDonald et al., 2000; Pieper et al., 2000; Wayman et al., 2001; Faro et al., 2002). However, their

PARP inhibitory effect *in vivo* is not determined solely by their potency *in vitro*, but most notably governed by their ability to cross cell membranes and their low lipophilicity. Thus, although different new chemical structures of potent PARP inhibitors have been discovered in the last decade (Southan and Szabó, 2003; Jagtap and Szabó, 2005), the need for developing selective inhibitors that are both potent and sufficiently water-soluble is still of pivotal importance (Woon and Threadgill, 2005).

In search of various chemical compounds as potential inhibitors for PARP-1, scientists from NYCOMED GmbH (Konstanz, Germany; formerly Altana Pharma AG) identified three new chemical entities that comprised imidazoquinolinone, imidazopyridine, or isoquinolindione structure, from which one compound, the isoquinolindione BYK204165 (Figure 5), displayed 100-fold selectivity for PARP-1, thus being a novel and valuable tool for investigating PARP-1 mediated effects. In collaboration with the Preclinical Research of NYCOMED GmbH (Konstanz, Germany) and scientists from William Harvey Research Institute (London, England), the PARP-1 selective inhibitor BYK204165 was investigated in comparison with other potent but unselective compounds for inhibition of (i) cell-free recombinant hPARP-1 and murine PARP-2 (mPARP-2), (ii) poly(ADP-ribose) synthesis in human lung epithel A549 and cervical carcinoma C4I cells as well in rat cardiac myoblast H9c2 cells after PARP activation by H_2O_2, and (iii) infarct size caused by coronary occlusion and reperfusion in anesthetized rats. These studies were completed by own experiments for inhibition of PAR formation in $Parp-1^{+/+}$ and $Parp-1^{-/-}$ mouse fibroblasts by the PARP-1 selective BYK204165, in comparison with the PARP-1/PARP-2 unselective BYK236864 (Figure 5), whereby nuclear PAR synthesis by both PARP-1 and PARP-2 was visualized and differentiated by immunofluorescence. Part of these results have already been published during work on this thesis (Eltze *et al.*, 2008) and further results are described in chapter 4.1.1.

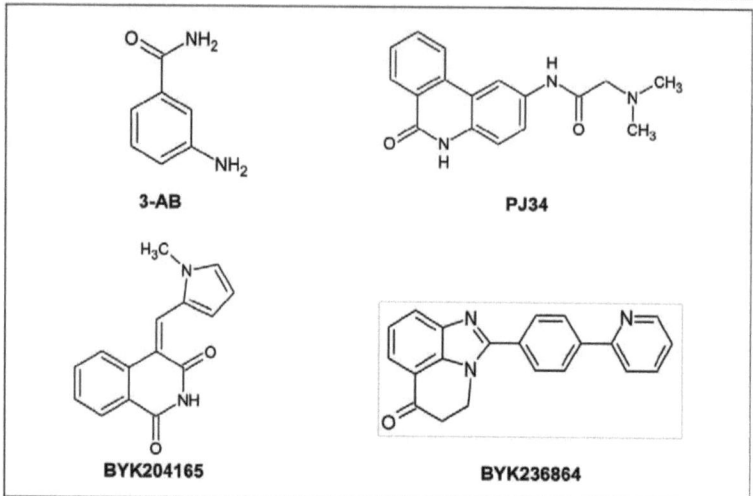

Figure 5: Chemical structures of 3-AB, PJ34, BYK204165 and BYK236864. Except for BYK236864, all inhibitors bear the benzamide structure.

2 Objective

In the past decades, a lot of cell-based studies as well as several knock-out mice have highlighted the role of PARP-1 as a key factor in genomic stability under conditions of genotoxic stress and provided new insights in its major role in DNA repair, transcription regulation and the recovery of cells after DNA damage.

As the participation of PARP-1, PARP-2 or both in many cellular processes often remains undistinguishable, mostly due to the lack of sufficiently selective inhibitors, the necessity for the design of isoform-selective inhibitors would be of great advantage to specifically study the role of PARP-1. Using 3T3 fibroblast from $Parp\text{-}1^{+/+}$ and $Parp\text{-}1^{-/-}$ mice that express both isoforms or only PARP-2, respectively, the potency and 100-fold selectivity of a new PARP-1 inhibitor, BYK204165, previously only being characterized on recombinant hPARP-1 and mPARP-2, should be tested in cellular systems.

To get further insights into the involvement of hPARP-1 and the consequences of its overexpression for cellular integrity, a Chinese hamster ovary (CHO) cell line overexpressing hPARP-1 (COMF10) was exposed to alkylating agents and X-irradiation, in order to pursue the processes of DNA repair and viability, or cell death via apoptosis or necrosis. Moreover, measurements of DNA repair kinetics in hPARP-1 overexpressing mouse lymphoma cells (EL-4) should clarify the role of PARP-1 in repairing DNA damage after X-irradiation in the presence or absence of pharmacological PARP inhibition.

Up to date, there have been no investigations made on the overexpression of PARP-1 in a transgenic mouse model. Most recently however, successful efforts of our group led to the generation of a mouse model with duplication of *mParp-1* flanking gene loci and moderate ectopic expression of hPARP-1 in addition to endogenous mPARP-1 (Mangerich *et al.*, 2009). These mice exhibit a moderate, multifaceted pathological phenotype showing signs of premature aging that could be related to increased levels of inflammation (Mangerich *et al.*, manuscript in preparation). On the other hand, a positive correlation of cellular poly(ADP-ribosyl)ation capacity and life span has been demonstrated in leukocytes of 13 mammalian species, whereby the longest-lived species studied (human) displayed 5-fold the level of maximal PARP activity of the shortest-lived (rat), while PARP-1 protein levels in the species did not correlate (Grube and Bürkle, 1992). Viewed together, results from a variety of experimental approaches and biological systems convincingly provide support for the notion that longevity is linked with high poly(ADP-ribosyl)ation capacity. This is supported by a recent study showing that *Parp-1* knock-out mice age moderately faster compared to wt controls (Piskunova *et al.*, 2008). However, the precise functions of PARP in controlling the process of aging *in vivo* have not been fully elucidated yet. To clarify the biological

Objective

consequences of hPARP-1 overexpression *in vivo*, that might also influence the process of cellular aging, the third objective was to generate *hPARP-1* transgenic mice with a respective protein overexpression in T-cells. This approach was motivated by the fact that the immune system interacts with every tissue and organ, but specially renders T-cells most suitable for such studies, as they are profoundly affected by aging. For this purpose, founder mice comprising a transgene for hPARP-1 under the control of a strong promoter should be generated, and then mated with mice supplied with Cre recombinase expression. The *Cre-lox* recombination system should provide a useful tool to specifically overexpress hPARP-1 in T-cells and also in other tissues, as different tissue-specific Cre recombinase expressing mice are commercially available.

3 Material and methods

3.1 Material

3.1.1 Chemicals and reagents

Table 1: Chemicals and reagents

Substance	Source
2-Mercaptoethanol	Sigma-Aldrich, Steinheim, Germany
2-Propanol	Riedel-de-Häen, Seelze, Germany
4-(2-Hydroxyethyl)-1-piperazineethanesulfonic acid (HEPES)	Roth, Karlsruhe, Germany
4′,6′-Diamidino-2-phenylindole (DAPI)	Sigma-Aldrich, Steinheim, Germany
Acetic acid (100%)	VWR, Darmstadt, Germany
Agarose (SeaKEM)	Biozym, Hess Oldendorf, Germany
Ammonium persulfate (APS)	Serva, Heidelberg, Germany
Ampicillin	Sigma-Aldrich, Steinheim, Germany
Annexin V	Alexis/Enzo Life Sciences, Lörrach, Germany
Aqua-Polymount	Polysciences, Eppelheim, Germany
Bacto agar	Becton-Dickinson, Heidelberg, Germany
Bacto trypton	Becton-Dickinson, Heidelberg, Germany
Bacto yeast extract	Becton-Dickinson, Heidelberg, Germany
Biotinylated SDS-PAGE standard broad range	Bio-Rad, München, Germany
Bovine serum albumine	Sigma-Aldrich, Steinheim, Germany
Bromphenol blue	Sigma-Aldrich, Steinheim, Germany
BYK204165 (4-(1-methyl-1H-pyrrol-2-ylmethylene)-4H-isoquinolin-1,3-dione)	Nycomed GmbH, Konstanz, Germany
BYK236864 (2-(4-pyridin-2-ylphenyl)-4,5-dihydro-imidazo[4,5,1-i,j]quinolin-6-one)	Nycomed GmbH, Konstanz, Germany
CasyClean	Schärfe System, Reutlingen, Germany
CasyTon	Schärfe System, Reutlingen, Germany
Chloroform	Merck, Darmstadt, Germany
Complete protease inhibitor cocktail	Roche Diagnostics, Mannheim, Germany
Cyclohexanediamine tetraacetate	Sigma-Aldrich, Steinheim, Germany
Dexamethasone	Sigma-Aldrich, Steinheim, Germany
Dimethyl sulfoxide (DMSO)	Sigma-Aldrich, Steinheim, Germany

Material and methods

Dithiothreitol (DTT)	Sigma-Aldrich, Steinheim, Germany
DNAse I	Roche, Mannheim, Germany
Dulbecco's modified Eagle medium (DMEM), high glucose	Gibco, Invitrogen, Karlsruhe, Germany
Ethanol (99.8%)	Riedel-de-Häen, Seelze, Germany
Ethidium bromide (10 mg/ml)	Sigma-Aldrich, Steinheim, Germany
Ethylenediamine-tetraacetic acid disodium salt dihydrate (EDTA)	Roth, Karlsruhe, Germany
Fetal bovine serum (FBS)	Gibco, Invitrogen, Karlsruhe, Germany
Fetal calf serum (FCS)	Biochrom, Berlin, Germany
Ficoll solution	Sigma-Aldrich, Steinheim, Germany
Formaldehyde (37%)	Riedel-de-Häen, Seelze, Germany
Geneticin disulfate (G418)	PAA, Pasching, Austria
Glucose	Merck, Darmstadt, Germany
Glycerol	Acros, Geel, Belgium
Glycine	Roth, Karlsruhe, Germany
Guanidine hydrochloride	Sigma-Aldrich, Steinheim, Germany
Hoechst 33342	Molecular Probes, Leiden, The Netherlands
Hydrochloric acid (37%)	Riedel-de-Häen, Seelze, Germany
Hydrogen peroxide (H_2O_2)	Sigma-Aldrich, Steinheim, Germany
Hygromycin B	Calbiochem, Darmstadt, Germany
Isoamyl alcohol	Merck, Darmstadt, Germany
JetPEI™ transfection reagent	Polyplus transfection, Illkirch, France
KH_2PO_4	Riedel-de-Häen, Seelze, Germany
L-Glutamine (100×)	Gibco, Invitrogen, Karlsruhe, Germany
Luminol	Fluka, Buchs, Switzerland
Magnesium chloride ($MgCl_2$)	Gibco, Invitrogen, Karlsruhe, Germany
Magnesium sulfate ($MgSO_4$)	Merck, Darmstadt, Germany
Manganese chloride ($MnCl_2$)	Merck, Darmstadt, Germany
meso-Inositol	Sigma-Aldrich, Steinheim, Germany
Methanol	Riedel-de-Häen, Seelze, Germany
Methyl methanesulfonate (MMS)	Sigma-Aldrich, Steinheim, Germany
MilliQ water	Millipore, Schwalbach, Germany
N,N,N',N'-tetramethylethylenediamine (TEMED)	Serva, Heidelberg, Germany

Material and methods

Na$_2$HPO$_4$	Riedel-de-Häen, Seelze, Germany
Na$_2$HPO$_4$·2H$_2$O	Riedel-de-Häen, Seelze, Germany
N-Methyl-N´-nitro-N-nitrosoguanidine (MNNG)	Sigma-Aldrich, Steinheim, Germany
Oligo(dt) primer	Promega, Mannheim, Germany
PageRuler Prestained Protein Ladder	MBI-Fermentas, St. Leon-Rot, Germany
Paraformaldehyde	Serva, Heidelberg, Germany
Penicillin/streptomycin (100×)	Gibco, Invitrogen, Karlsruhe, Germany
Phenol	Roth, Karlsruhe, Germany
PIPES (piperazine-N,N'-bis(ethanesulfonic acid)	Sigma-Aldrich, Steinheim, Germany
PJ34 (N-(6-oxo-5,6-dihydro-phenanthridin-2-yl)-N,N-dimethylacetamide)	Alexis/Enzo Life Sciences, Lörrach, Germany
Poly-L-lysine	Sigma-Aldrich, Steinheim, Germany
Propidium iodide	Sigma-Aldrich, Steinheim, Germany
Ribonuclease inhibitor (40 U/µl)	MBI Fermentas, St. Leon-Rot, Germany
Rotiphorese 30% acrylamide/bisacrylamide (37.5:1)	Roth, Karlsruhe, Germany
RPMI medium 1640	Gibco, Invitrogen, Karlsruhe, Germany
Skim milk powder	Rapilait, Sulgen, Schwitzerland
Sodium acetate	Merck, Darmstadt, Germany
Sodium azide (NaN$_3$)	Merck, Darmstadt, Germany
Sodium pyruvate (100×)	Gibco, Invitrogen, Karlsruhe, Germany
Sybr®Green (10000x concentrated)	Mo Bi Tec, Göttingen, Germany
Sytox®	MolecularProbes, Leiden, The Netherlands
Triton X-100	Sigma-Aldrich, Steinheim, Germany
Trizma base	Sigma-Aldrich, Steinheim, Germany
Trypsin/EDTA (0.25%/1 mM)	Gibco, Invitrogen, Karlsruhe, Germany
Tween 20	Sigma-Aldrich, Steinheim, Germany
Urea	Merck, Darmstadt, Germany

Material and methods

3.1.2 Laboratory equipment

Table 2: Laboratory equipment

Description/Specification	Manufacturer
Cell counter, Casy Model TT	Schärfe System, Reutlingen, Germany
Cell culture material, dishes, flasks, pipettes	Corning, Schiphol-Rijk, The Netherlands
Centrifuges: Beckman Coulter LE-80K Biofuge Fresco Biofuge Pico Function Line Labofuge 400 Megafuge 1.0R Centrifuge 5415 R Centrifuge 5810 R	 Beckman Coulter, Krefeld, Germany Heraeus, Fellbach, Germany Heraeus, Fellbach, Germany Heraeus, Fellbach, Germany Heraeus, Fellbach, Germany Eppendorf, Hamburg, Germany Eppendorf, Hamburg, Germany
Chemiluminescence detector, Image Reader LAS-1000 Pro	Fujifilm, Düsseldorf, Germany
Cooling water thermostat, Lauda E 200, Ecoline RE 204	Lauda Dr. R. Wobser GmbH, Lauda-Königshofen, Germany
Cryo freezer 1°C (Mr. Frosty)	Wessington Cryogenics, Houghton-le-Spring, UK
Cytospin	Heraeus, Fellbach, Germany
Dark box, Intelligent Dark Box	Fujifilm, Düsseldorf, Germany
ELISA reader, SLT Spectra	SLT Labinstruments, Crailsheim, Germany
Flow cytometry device, BD LSRII	BD Biosciences, Heidelberg, Germany
Fluorescence microscopes, Axiovert 200M equipped with an AxioCam MRm	Zeiss, Göttingen, Germany
Gel documentation, UV Systeme	Intas, Göttingen, Germany
Incubator for mammalian cell culture Hera Cell 240	Heraeus, Fellbach, Germany
Incubators for bacterial cell cultures: Infors HT, Minitron	Infors, Bottmingen, Switzerland
Microplate fluorescence reader FL600	BIO-TEK, Bad Friedrichshall, Germany
Microscope slides, SuperFrost	Menzel, Braunschweig, Germany
Millipore Millex-GV 0.22 µm filter	Millipore, Schwalbach, Germany
Millipore membrane VMW PO2500/VM 0.05 µm	Millipore, Schwalbach, Germany
Mouse housing: isolated ventilated cage unit (IVC)	Techniplast, Hohenspeißenberg, Germany
Neubauer chamber	Superior, Lauda-Königshofen, Germany
p-Coumaric acid	Fluka, Buchs, Switzerland

Material and methods

PCR Cycler:	
Mastercycler Gradient	Eppendorf, Hamburg, Germany
MultiCycler PTC 200	Bio-Rad, Munich, Germany
iCycler	Bio-Rad, Munich, Germany
Photometers :	
BioPhotometer	Eppendorf, Hamburg, Germany
Ultraspec 2100 Pro	Amersham Biosciences, Freiburg, Germany
NanoDrop 1000	Thermo Scientific, Schwerte, Germany
SDS PAGE unit, Hoefer MiniVE system	Amersham Biosciences, Freiburg, Germany
TECAN robot, Genesis RSP 100	TECAN AG, Hombrechtikon, Switzerland
Thermomixer comfort	Eppendorf, Hamburg, Germany
Transfer membrane, protein, Hybond-ECL nitrocellulose membrane	Amersham Biosciences, Freiburg, Germany
Ultrasound device, Sonorex TK 52	Bandelin, Berlin, Germany
Western blot wet transfer unit, Hoefer Mini Blot Module	Amersham Biosciences, Freiburg, Germany
X-ray dosimeter, PTW unidos E	PTW-Freiburg, Freiburg, Germany
X-ray equipment, RT 100	C.H.F. Müller GmbH, Hamburg, Germany
X-ray tube, TÖ 100/8 („Oberflächenröhre")	C.H.F. Müller GmbH, Hamburg, Germany

3.1.3 Buffers and solutions

Table 3: Buffers and solutions

Designation	Composition
2YT Medium	Difco bacto yeast extract 50 g, difco bacto tryptone 80 g, NaCl 25 g, H_2O ad 5000 ml, autoclaved
Alkylating unwinding buffer	42.5% Lysis buffer in 0.2 M NaOH
Annexin binding buffer	10 mM HEPES, 140 mM NaCl, 2.5 mM $CaCl_2$, pH 7.4
Blocking solution	5% (w/v) skim milk powder in TNT buffer
COMF10 cell medium	DMEM with pyruvate, 100 U/ml penicillin, 100 µg/ml streptomycin, 800 U/ml hygromycin B, 800 µg/ml geneticin, 10% (v/v) heat-inactivated FCS
Complete protease inhibitor cocktail	1×-concentrated in PBS
COPF5 cell medium	DMEM with pyruvate, 100 U/ml penicillin, 100 µg/ml streptomycin, 800 U/ml hygromycin B, 800 µg/ml geneticin, 2 mM L-glutamine, 10% (v/v) heat-inactivated FCS
COR4 cell medium	DMEM with pyruvate, 100 U/ml penicillin, 100 µg/ml streptomycin, 800 U/ml hygromycin B, 10% (v/v) heat-inactivated FCS

Material and methods

Dex stimulation medium	DMEM with pyruvate, 100 nM dexamethasone, 100 U/ml penicillin, 100 µg/ml streptomycin, 2 mM L-glutamine, 10% (v/v) heat-inactivated FCS
EL-4 cell medium	DMEM w/o pyruvate, 100 U/ml penicillin, 100 µg/ml streptomycin, 2 mM L-glutamine 10% (v/v) heat-inactivated FCS
Enhanced chemiluminescence (ECL) solution	Solution 1: luminol (250 mM in DMSO) 50 µl, p-coumaric acid (90 mM in DMSO) 22 µl, Tris-HCl (1 M, pH 8.5) 0.5 ml, MilliQ H_2O 4.4 ml Solution 2: H_2O_2 (30%) (v/v) 3 µl, Tris-HCl (1 M, pH 8.5) 0.5 ml, MilliQ H_2O 4.4 ml Mix of solution 1+2 in equal amounts before use
Fixing solution	4% (v/v) Formaldehyde in PBS
Flow cytometry buffer	0.5% (v/v) FBS, 2 mM NaN_3, in PBS
Freezing medium	90% FCS, 10% DMSO (v/v)
Glycine solution	100 mM Glycine in PBS
HEK293T cell medium, 3T3 cell medium	DMEM with pyruvate, 100 U/ml penicillin, 100 µg/ml streptomycin, 2 mM L-glutamine, 10% (v/v) heat-inactivated FCS
LB medium (lysogeny broth)	Difco bacto yeast extract 5 g, difco bacto tryptone 10 g, 5 g NaCl, add 1 l MilliQ water, pH 7.0, autoclaved
Lysis buffer	9 M Urea, 2.5 mM cyclohexanediamine tetraacetate, 0.1% SDS
Microinjection buffer	10 mM Tris-HCl (pH 7.5), 0.1 mM EDTA, ad H_2O (Millipore quality), 0.2 µm pore size filtrated, autoclaved
Neutralisation buffer	1 M Glucose, 14 mM 2-mercaptoethanol
PBS (phosphate-buffered saline)	137 mM NaCl, 10 mM Na_2HPO_4, 3 mM KH_2PO_4, pH 7.4, autoclaved
Permeabilization solution	0.4% (v/v) Triton X-100 in PBS
SDS-PAGE 1.5× high-urea protein loading buffer	93.75 mM Tris-HCl (pH 6.8), 9 M urea, 7.5% (v/v) 2-mercaptoethanol, 15% (v/v) glycerol, 3% (w/v) SDS, 0.01% (w/v) bromphenol blue
SDS-PAGE 2x sample buffer	1.25 ml Tris-HCl (pH 6.7), 10% (w/v) SDS, 0.8% (v/v) 2-mercaptoethanol, 5% (v/v) glycerol, 0.01% (w/v) bromphenol blue, ad H_2O to 10 ml
SDS-PAGE Laemmli buffer	25 mM Tris-HCl (pH 8.6), 192 mM glycine, 0.1% (w/v) SDS
SDS-PAGE resolving gel buffer	3 M Tris-HCl (pH 8.9)
SDS-PAGE stacking gel buffer	0.5 M Tris-HCl (pH 6.7)
SOC medium	2% (w/v) Bacto trypton, 0.5% (w/v) bacto yeast extract, 10 mM NaCl, 2.5 mM KCl, 10 mM $MgCl_2$, 10 mM $MgSO_4$, 20 mM glucose, autoclaved

Suspension buffer	0.25 M meso-Inositol, 10 mM $Na_2HPO_4 \cdot 2H_2O$, 1 mM $MgCl_2$, pH 7.4
Sytox®/Hoechst mix	50 µl Sytox® (250 µM), 20 µl Hoechst 33342 (2.5 mg/ml), 180 µl DMSO
TB buffer	10 mM PIPES, 15 mM $CaCl_2$, 250 mM KCl, pH adjusted to 6.7 with KOH, addition of 55 mM $MnCl_2$
TNT buffer	150 mM NaCl, 10 mM Tris-HCl pH 8.0, 0.05% (v/v) Tween 20
Towin buffer	50 mM Tris-HCl pH 8.6, 384 mM glycine, 20% (v/v) methanol, 0.1% (w/v) SDS
Tris acetate EDTA (TAE) buffer	40 mM Tris-HCl (pH 8.0), 20 mM acetic acid, 1 mM EDTA

All buffer solutions were prepared in MilliQ water

3.1.4 Plasmids

Table 4: Plasmids

Designation	Source
P1017	Roger M. Perlmutter, University of Washington, Seattle, USA
PGKneotpAlox2	Phil Soriano, Fred Hutchinson Cancer Research Center, Seattle, USA
PGKneotpAlox2-NotI	Present thesis
pJD5	Dr. Jörg Diefenbach, University of Konstanz, Germany
pMD2.G	Addgene, Cambridge, USA
pPARP25	Küpper J. H., Doctoral thesis, University of Heidelberg, 1990
pPARP31 (*hPARP-1* cDNA)	Van Gool et al. (Van Gool et al., 1997)
pSL1180-*hPARP-1*	Dr. Sascha Beneke, University of Konstanz, Germany
pspPAX.2	Addgene, Cambridge, USA
pUC18	MBI-Fermentas, St. Leon-Rot, Germany
pUCTE1	Present thesis
pUCTE2	Present thesis
pUCTE3	Present thesis
pUCTE4	Present thesis

	Material and methods
pUCTE5	Present thesis
pWPT-*GFP*	Addgene, Cambridge, USA
pWPT-*hPARP-1*	Present thesis
Turbo-*Cre*	Dr. Timothy Ley, University of Washington, USA
Ubi-junB	M. Schorpp (Schorpp et al., 1996)

3.1.5 Oligonucleotides

Table 5: Oligonucleotides

Name	Description	Sequence 5′-3′	Orientation
AMa19	Genotyping *Neo*	TGCTCCTGCCGAGAAAGTATCCATCATGGC	Sense
AMa20	Genotyping *Neo*	CGCCAAGCTCTTCAGCAATATCACGGGTAG	Antisense
AMa21	Genotyping *Cre*	GCATTACCGGTCGATGCAACGAGTGATGAG	Sense
AMa22	Genotyping *Cre*	GAGTGAACGAACCTGGTCGAAATCAGTGCG	Antisense
AMa25	Genotyping *Fabpi* 200 (fatty acid binding protein, intestinal)	TGGACAGGACTGGACCTCTGCTTTCCTAGA	Sense
AMa26	Genotyping *Fabpi* 200	TAGAGCTTTGCCACATCACAGGTCATTCAG	Antisense
AMa09	qPCR reference gene, murine and human *Cygb*	CAACACTGTCGTGGAGAACC	Sense
AMa10	qPCR reference gene, murine and human *Cygb*	GGTTCCACCTTGTGCTTGAG	Antisense
30302	Genotyping *hPARP-1* cDNA	ATGGTGTAGACGTTCCTCTTGGGACC	Sense
30401	Genotyping *hPARP-1* cDNA	ATGGCGGAGTCTTCGGATAAGCTC	Sense
30402	Genotyping *hPARP-1* cDNA	AGCCACAGCTAGGCATGATTGACC	Sense
30403	Genotyping *hPARP-1* cDNA	ATGGTGTTCGGTGCCCTCCTTCCC	Sense
30404	Genotyping *hPARP-1* cDNA	GAGTTGTGTCTGAGGACTTCCTCC	Sense
30405	Genotyping *hPARP-1* cDNA	TTCCCCAAGGGCATCTTCTGAAGG	Antisense
30406	Genotyping *hPARP-1* cDNA	ACCCGTGCCACAGCAATCTTCGG	Antisense
30407	Genotyping *hPARP-1* cDNA	ACTCGGCTACCTCTCCCAATTACC	Antisense
30701	Genotyping hPARP-1 cDNA	TGTGGAGGGCGGAGGCGTGG	Antisense

Material and methods

40326	Real-time PCR, binds to *hPARP-1* cDNA, HEX labelled TaqMan probe, used in combination with primer pair 40804/40805	HEX-TGTGGAGGGCGGAGGCGTGG-BHQ1	Antisense
40327	Real-time PCR, binds to murine β-actin, used in combination with probe 40329	TGCGTGACATCAAAGAGAAG	Sense
40328	Real-time PCR, binds to murine β-actin, used in combination with probe 40329	CAGCTCATAGCTCTTCTCC	Antisense
40329	Real-time PCR, binds to β-actin, FAM labelled TaqMan probe, used in combination with primer pair 40327/40328	FAM-ACTGCCGCATCCTCTTCCTCCC-BHQ1	Sense
40804	Real-time PCR, binds to murine and *hPARP-1* cDNA, used in combination with probe 40326	YCCAGAAASCAGCGCC	Sense
40805	Real-time PCR, binds to murine and *hPARP-1* cDNA, used in combination with probe 40326	GCTTCCCYAGAGTCAGG	Antisense
50303	Multiple cloning site (MCS), phosphorylated	AATTCCGTACGAGATCTGTCGACTCTAGAGAGCTCACGCGTCTCGAGCTGCAGCCCGGGATCGATGCGGCCGCCGTACGA	Sense
50304	Multiple cloning site (MCS), phosphorylated	AGCTTCGTACGGCGGCCGCATCGATCCCGGGCTGCAGCTCGAGACGCGTGAGCTCTCTAGAGTCGACAGATCTGTACGG	Antisense
50402	Genotyping *hGh* minigene	TGTGCCAAAGGGATTTTAGG	Antisense
50404	Genotyping *hUbiC* promoter	GCTGTGAGGTCGTTGAAACA	Sense
50405	Genotyping Stop sequence	GGTTCCGGATCCACTAGTTCT	Antisense
50509	qPCR, *Neo*	GTTGTCACTGAAGCGGGAAG	Sense
50510	qPCR, *Neo*	GGATACTTTCTCGGCAGGAG	Antisense
50602	Genotyping *hPARP-1* cDNA	GCTCCCAGGAGTCAAGAGTG	Sense
50603	Genotyping *hPARP-1* cDNA	CAACTCCTGAAGGCTCTTGG	Antisense
50610	Flanking PCR	TTGGGTCAATATGTAATTTTCAGTGT	Sense
50611	Genotyping *hPARP-1* cDNA, flanking PCR	GCCCTTTTCTATCTTCTCCATACA	Antisense

Oligonucleotides were purchased from Invitrogen, Karlsruhe, Germany or from MWG, Ebersberg, Germany

3.1.6 PCR conditions

Table 6: PCR conditions

PCR	Primers	Amplicon size	Polymerase	Vol.	Conditions
Cre/ control PCR	AMa 21/22 (*Cre*) AMa 25/26 (*Fabpi*)	408 bp 200 bp	0.5 U HotMaster Taq	25 µl	94°C 2 min, 62°C 20 s, 68°C 1 min, (33 cycles of 94° 20 s, 62°C 20 s, 68°C 1 min), 68°C 10 min
NeoR/ control PCR	AMa 19/20 (*Neo*) AMa 25/26 (*Fabpi*)	380 bp 200 bp	0.5 U HotMaster Taq	10 µl	94°C 2 min, 62°C 20 s, 68°C 1 min, (33 cycles of 94° 20 s, 62°C 20 s, 68°C 1 min), 68°C 10 min
hPARP/ actin PCR	30404/30405 (*hPARP*) 40327/40328 (*β-actin*)	714 bp 100 bp	0.5 U HotMaster Taq	10 µl	95°C 2 min, 59°C 20 s, 68°C 1 min, (35 cycles of 95°C 20 s, 59°C 20 s, 68°C 1 min), 68°C 10 min
qPCR, transgene copy number	50509/50510 (*Neo*) AMa 09/10 (*CytB*)	76 bp 95 bp	0.6 U iTaq DNA polymerase	25 µl	95°C 3 min, (50 cycles of 95°C 15 s, 60 °C 1min), 95°C 1 min, 50°C 1 min, (90 cycles of 0.5°C/cycle increments, 10 s each)
Flanking PCR, *Neo*/Stop	50610/50611 (*Neo*/Stop)	3400 bp (+*Neo*/Stop) 650 bp (-*Neo*/Stop)	0.2 U KOD Hot Start DNA Polymerase	10 µl	95°C 2min, (35 cycles of 95°C 20 s, 57°C 10 s, 70°C 1 min 40 s)
cDNA PCR, hPARP-1	30401/30701 (hPARP-1)	1120 bp	0.4 U KOD Hot Start DNA Polymerase	20 µl	95°C 2 min, (13 cycles of 95°C 20 s, 64°C 10 s, decrement 0.5°C each cycle, 70°C 25 s), (23 cycles of 95°C 20 s, 61.5°C 10 s 70°C 25 s)
Meltcurve	-	-	-	-	95°C 1 min, 50°C 1 min, 90 cycles of 0.5°C increment, each cycle 10 s

3.1.7 Molecular weight standards

Table 7: DNA and protein ladders

Designation	Manufacturer
100bp ladder	New England Biolabs, Frankfurt, Germany

Material and methods

Biotinylated SDS-PAGE standard, broad-range	Bio-Rad, Munich, Germany
GeneRuler 1kbp ladder	MBI Fermentas, St. Leon-Rot, Germany
GeneRuler DNA ladder mix	MBI Fermentas, St. Leon-Rot, Germany
Low molecular weight DNA ladder	New England Biolabs, Frankfurt, Germany
MassRuler DNA ladder mix	MBI Fermentas, St. Leon-Rot, Germany
PageRuler prestained protein ladder	MBI Fermentas, St. Leon-Rot, Germany

3.1.8 Kits

Table 8: Kits

Kit	Manufacturer
Direct PCR-Tail	peqLab Biotechnologie, Erlangen, Germany
EndoFree plasmid maxi kit	Qiagen, Hilden, Germany
High pure PCR template preparation kit	Roche Diagnostics, Mannheim, Germany
MACS® Pan T-cell isolation kit	Miltenyi Biotec, Bergisch Gladbach, Germany
MiniElute gel extraction kit	Qiagen, Hilden, Germany
MiniElute reaction cleanup kit	Qiagen, Hilden, Germany
Omniscript RT kit	Qiagen, Hilden, Germany
peqGOLD TriFast	peqLab Biotechnologie, Erlangen, Germany
PhoenIX Maxiprep kit	MP Biomedicals, Heidelberg, Germany
Qiagen plasmid giga kit	Qiagen, Hilden, Germany
QIAprep Spin Miniprep kit	Qiagen, Hilden, Germany
RNase-Free DNase set	Qiagen, Hilden, Germany
RNeasy Mini kit	Qiagen, Hilden, Germany

3.1.9 Enzymes

Table 9: Enzymes

Name	Manufacturer
Antarctic phosphatase (5000 U/ml)	New England Biolabs, Frankfurt, Germany
DNase I	Roche Diagnostics, Mannheim, Germany
Mung Bean nuclease	New England Biolabs, Frankfurt, Germany
Proteinase K	Sigma-Aldrich, Steinheim, Germany
Ready-To-Go T4 DNA ligase	Amersham Biosciences, Freiburg, Germany

3.1.10 Restriction enzymes

Table 10: Restriction enzymes

Name	Manufacturer
*Bam*HI	New England Biolabs, Frankfurt, Germany
*Bcl*I	MBI Fermentas, St. Leon-Rot, Germany
*Bgl*II	MBI Fermentas, St. Leon-Rot, Germany
*Eco*RI	New England Biolabs, Frankfurt, Germany
*Hin*dIII	MBI Fermentas, St. Leon-Rot, Germany
*Mlu*I	MBI Fermentas, St. Leon-Rot, Germany
*Not*I	MBI Fermentas, St. Leon-Rot, Germany
*Pfl*23II	MBI Fermentas, St. Leon-Rot, Germany
*Pst*I	New England Biolabs, Frankfurt, Germany
*Sac*I	MBI Fermentas, St. Leon-Rot, Germany
*Sma*I	MBI Fermentas, St. Leon-Rot, Germany
*Xba*I	MBI Fermentas, St. Leon-Rot, Germany
*Xho*I	New England Biolabs, Frankfurt, Germany

3.1.11 Polymerases

Table 11: Polymerases

Designation	Manufacturer
High Fidelity (HiFi) PCR enzyme mix (5 U/µl)	MBI Fermentas, St. Leon-Rot, Germany
HotMaster Taq DNA polymerase (5 U/µl)	VWR, Darmstadt, Germany
iQ Sybr Green Supermix, iTaq DNA polymerase (0.05 U/µl))	Bio-Rad, Munich, Germany
KOD Hot Start DNA polymerase	Novagen/Merck, Darmstadt, Germany
RealMasterMix Probe, HotMaster Taq DNA polymerase (0.1 U/µl)	Eppendorf, Hamburg, Germany
T4 DNA polymerase	New England Biolabs, Frankfurt, Germany

3.1.12 Antibodies

Table 12: Antibodies

Name	Description/Application	Manufacturer/Source
Anti-actin (MAB1501)	Monoclonal mouse antibody against β-actin (species unspecific)	Millipore/Chemicon, Schwalbach, Germany
Anti-PARP (CII-10)	Monoclonal mouse antibody against the N-terminal DNA-binding domain of PARP-1 (species unspecific)	Hybridoma cells from G. G. Poirier, Québec, Canada
Anti-human-PARP-1 (FI-23)	Monoclonal mouse antibody against the second zinc finger (hPARP-1 specific)	Hybridoma cells from G. G. Poirier, Québec, Canada
Anti-PAR (10H)	Monoclonal mouse antibody against poly(ADP-ribose)	Hybridoma cells from M. Miwa and T. Sugimura, Tokyo, Japan
Anti-Cre	Polyclonal rabbit antibody against Cre recombinase	Novagen/Merck, Darmstadt, Germany
Polyclonal goat anti-mouse IgG, horse-radish peroxidase (HRP)-conjugated	Secondary antibody	DakoCytomation, Hamburg, Germany
Polyclonal goat anti-rabbit IgG, HRP-conjugated	Secondary antibody	DakoCytomation, Hamburg, Germany
Polyclonal goat anti-mouse IgG, AlexaFluor 488-conjugated	Secondary antibody	MoBiTec Molecular Probes, Göttingen, Germany
Polyclonal goat anti-rabbit IgG, AlexaFluor 568-conjugated	Secondary antibody	MoBiTec Molecular Probes, Göttingen, Germany

3.1.13 Cell lines

Table 13: Cell lines

Name	Description
HEK293T	Human epithelial kidney cells, transformed by simian virus 40 (SV40) large T antigen, semi-adherent; ATCC number: CRL-11268
CO60	SV40-transformed Chinese hamster ovary (CHO) cells; S. Lavi, Tel Aviv, Israel
COR4	Constitutive expression of human glucocorticoid receptor encoded by plasmid pHG0 and hygromycin B resistance plasmid pTKhygro; P. Chambon and H. Gronemeyer, Illkirch-Graffenstaden, France
COMF10	Stable super-transfection of COR4 cells with

Material and methods

	plasmid pPARP93, encoding full-length *hPARP-1* cDNA under transcriptional control of mouse mammary tumor virus (MMTV) long terminal repeat promoter and *pTKneo* (geneticin-resistance) (Meyer *et al.*, 2000); R. Meyer, DKFZ Heidelberg, Germany
EL-4	Mouse T lymphocyte lymphoma cell line; ATTC number: TIB-39
3T3 (PARP-1 wt, PARP-1$^{-/-}$)	Mouse embryonic fibroblasts; G. de Murcia, Strasbourg, France
COPF5	Stable super-transfection of CO60 cells with plasmid pPARP93, encoding full-length *hPARP-1* cDNA under transcriptional control of human cytomegalovirus (*hCMV*) promoter and *pTKneo* (geneticin-resistance); Prof. A. Bürkle, University of Konstanz, Germany

3.1.14 Organisms

Table 14: Organisms

Organism	Source
Mus musculus (C57BL/6)	Charles River Laboratories, Sulzfeld, Germany
E. coli DH5α	Invitrogen, Karlsruhe, Germany
Mus musculus GMO (B6.FVB Tg (EIIa-Cre) C5379LMGD/JF)	Charles River Laboratories, Sulzfeld, Germany
Mus musculus B6.Cg-TgN(Lck-Cre)548Jxm	Charles River Laboratories, Sulzfeld, Germany

3.1.15 Software

Table 15: Software

Name	Manufacturer
ACD/ChemSketch 12	Advanced Chemistry Development, Toronto, Canada
AIDA Image Analysis Software 3.10	Raytest, Straubenhardt, Germany
AxioVision Rel. 4.6	Zeiss, Oberkochen, Germany
Clone Manager Suite 7	Sci Ed software, Cary, NC, USA
CorelDraw Graphics Suite X4	Corel, Ottawa, Canada
Endnote X2	Thomson ISI ResearchSoft, Carlsbad, CA, USA
Ensembl genome browser (http://www.ensembl.org)	European Bioinformatics Institute/Wellcome Trust Sanger Institute, Hinxton, UK

Material and methods

FlowJo 7.2.5	Tree Star, San Carlos, USA
Microsoft Office 2007	Microsoft, Redmond, USA
NCBI databases (http://www.ncbi.nlm.nih.gov)	National Center for Biotechnology Information, Bethesda, USA
Photoshop CS3 Extended	Adobe Systems, San Jose, CA, USA
Primer3 (http://frodo.wi.mit.edu)	Open source
Prism 5.02	GraphPad, San Diego, CA, USA
iCycler iQ Real-Time PCR Detection System Software Version 3.1	Bio-Rad Laboratories, Hercules, CA, USA
KC4™ v.3.3 for microplate fluorescence reading	BIO-TEK, Bad Friedrichshall, Germany

3.2 Methods

3.2.1 DNA repair measurement

3.2.1.1 Measurement of DNA strand breaks with FADU

To quantify the repair of DNA strand breaks after X-irradiation, a modified and automated version of the fluorimetric detection of alkaline DNA unwinding (FADU) method was used which is described in detail elsewhere (Moreno-Villanueva *et al.*, 2009). In brief, DNA strand breaks caused in cells after X-irradiation were detected by partial unwinding of double-stranded DNA under controlled alkaline conditions at damaged sites of the DNA. The unwinding was stopped by a neutralization buffer and thereafter, the amount of remaining double-stranded DNA was detected by the fluorescent probe Sybr®Green. In addition, two further samples were processed in parallel: 1) In order to quantify the total amount of DNA present, samples were treated with neutralisation buffer prior to the alkaline buffer. Thus, the critical alkaline pH necessary for DNA denaturation was not reached and unwinding did not occur. These samples were designated as T-samples. 2) Samples which were not X-irradiated but treated with alkaline buffer, underwent unwinding at the ends of the chromosomes, at endogenous DNA strand breaks and at replication forks, thus reflecting the physiological conditions of the living cell and were designated as P_0-samples (Moreno-Villanueva *et al.*, 2009). The experiments were performed as followed.

Cell cultures of mouse T-lymphocytes (EL-4) were grown to a cell density between 2×10^5 and 5×10^5 cells/ml in 75 cm² culture flasks filled with 20 ml DMEM medium and then centrifuged at 200 g for 5 min at room temperature (RT). The supernatant was removed and the cell pellet resuspended in 1 ml RPMI without FCS. The cell number was estimated in a Neubauer chamber and adjusted to 5×10^5 cells/ml for eight replicate measurements (2.5×10^5 cells/ml for quadruplicate measurements). Aliquots of 100 µl were transferred into 2 ml Eppendorf tubes, incubated at 37°C for 2 h for recovery and then X-irradiated with 7 Gy (70 keV energy, 1.25-mm aluminium filter, 9.4 mA current) on ice. For the detection of physiologically occurring DNA strand breaks and the amount of total DNA, a 100 µl aliquot of cell suspension for a P_0- and T-sample was transferred into a 2 ml Eppendorf tube and not X-irradiated. To allow DNA repair, cells were incubated for various time periods in a water bath at 37°C. Thereafter, the Eppendorf tubes were carried over into an ice-cold custom-made device mounted in the TECAN Genesis RSP 100 liquid handling working space. The immediate following steps were performed automatically and will be described in brief below (for detailed protocol, see Moreno-Villanueva *et al.*, 2009).

3.2.1.2 Automated FADU protocol

At first, 900 µl suspension buffer was added to 100 µl EL-4 cell suspension. From each Eppendorf tube, 70 µl/well of cell suspension was transferred to eight (or four) wells of a 96-well plate, followed by the addition of 70 µl lysis buffer per well. After incubation of 12 min, 70 µl/well of alkylating unwinding buffer was added slowly on the top of the cell lysate, whereas for T-samples 140 µl of neutralization buffer was added prior to the unwinding buffer. Cells were incubated for 15 min to allow diffusion of alkaline solution into the cell lysate. For alkaline unwinding of the DNA, the temperature of the samples was shifted from 0°C to 30°C and kept for further 60 min. To stop the DNA unwinding procedure, the temperature of the samples was reduced to 20°C and 140 µl of neutralization buffer (except for the T-samples) was added. In order to determine the amount of remaining double-stranded DNA, 156 µl of Sybr®Green (1:8,333 in H_2O; v/v) was added to each well, resulting in a total volume of 506 µl. The 96-well plates were manually transferred into a microplate fluorescence reader (SLT spectra) and after excitation at 485 nm the amount of emission was measured at 530 nm and quantified using the software KC4™.

3.2.2 DNA/RNA/protein isolation

3.2.2.1 Organ isolation from mice

Transgenic mice at the age of 1 and 12 months were euthanized by CO_2 inhalation in a closed cage. After fixation of the animal with pins, the fur was saturated with 70% ethanol and the skin was cut along the ventral midline from the caudal abdomen to the chin. Thereafter, the underlying muscle wall was incised along the ventral midline from the caudal abdomen to the sternum. Finally, the thorax was opened along the sternum and the various organs (thymus, spleen) were removed and snap-frozen in liquid nitrogen until use.

3.2.2.2 DNA isolation from organs

The DNA isolation from the taken organs was performed with peqGOLD TriFast™ (PeqLab) according to the manufacturer's instructions with the following options used: 50 –100 mg of frozen tissue were given in 1 ml TriFast™ solution and completely homogenized with a glass douncer (7 ml) at RT. Thereafter, the DNA was dissolved in 500 µl 8 mM NaOH and finally, for longtime storage at 4°C, pH was adjusted to pH 8.0 by adding 57.5 µl HEPES (0.1 M) and 2 µl of EDTA (0.5 M).

3.2.2.3 RNA isolation from organs

RNA isolation was performed either with peqGOLD TriFast™ (PeqLab) according to the manufacturer's instructions in the same way as described in section "DNA isolation from organs", or with the RNeasy Mini kit (Qiagen). To disrupt and homogenize the tissue, 30 mg of frozen tissue were given to 600 µl Buffer RLT (provided with RNeasy Mini kit) and homogenized by using a rotor-stator homogenizer at maximum speed for approximately 1 min until homogeneity. Subsequently, all steps were performed according to the manufacturer's instructions. For the complete elimination of DNA contamination, on-column DNase digestion with the RNase-free DNase set (Qiagen) was performed according to the standard protocols. RNA was eluted in 50 µl RNase-free water (Qiagen) and stored at -20°C until further use.

3.2.2.4 Protein isolation from organs

The RNA isolation was performed with peqGOLD TriFast™ (PeqLab) according to the manufacturer's instructions similarly as described in section "DNA isolation from organs". For solubilization of the protein pellet, it was incubated for 1 h in 1 ml 10 M urea/50 mM DTT at RT and then heated for 3 min at 99°C in a thermomixer (Eppendorf). Finally, the pellet was treated with ultrasound for 3 min at RT and completely dissolved by pipetting the resulting solution several times up and down. The protein concentration was quantified by photometric analysis following the equation:

Protein concentration (mg/ml) = $(A_{235} - A_{280})/2.51$ (Whitaker and Granum, 1980)

3.2.3 Isolation and purification of T-cells from thymus and spleen

To achieve a maximum of T-cell purity by removing non-T-cells in the thymus or spleen, the organ was treated prior to a later protein determination as follows: firstly, the freshly isolated organ was placed on a custom-made fine metal strainer (2x2 cm) in a 12 cm cell culture dish filled with 5 ml PBS. Then, with the backside of a plunger from a syringe the organ was pressed through this strainer. Residual small organ clumps were aspirated with a 5 ml sterile pipette and blown out several times through the strainer with maximal intensity until a uniform homogenous solution was achieved. Secondly, the cell solution had to be layered under 5 ml Ficoll solution in a 15 ml falcon tube and centrifuged at 860 g for 15 min at 12°C. Thirdly, the appropriate cells were sucked off with a glass Pasteur capillary pipette (approx. 2 ml) and passed through a 30 µm nylon mesh to avoid cell clumps. Fourthly, cells were counted and centrifuged at 300 g for 4 min at RT. All following steps were made with the MACS® Pan T-cell isolation kit (Miltenyi Biotec) beginning at step 2.2 "magnetic labeling" with the use of a MS column according to the manufacturer's instructions. Finally, the purity of T-cells was evaluated by flow cytometry with a fluorochrome-conjugated antibody against the pan T-cell marker CD90 as recommended by the manufacturer. The purified cells were stored at -20°C until use for Western blot analysis.

3.2.4 Genotyping of mice

To obtain DNA for genotyping of putative transgenic mice, tail tip biopsies of 0.2 - 0.5 cm length were taken from mice at the age of six weeks under sterile conditions and frozen at -20°C until further use. DNA was isolated by using the DirectPCR® lysis reagent (PeqLab) according to the manufacturer's instructions. The lysates obtained were stored at -20°C until use.

3.2.5 Determination of transgene copy number

The transgene copy number in the 18 different transgenic founder lines were determined by quantitative real-time PCR (qPCR) using the nonspecific intercalating dye Sybr Green I (iQ Sybr Green Supermix, Bio-Rad) for accurate detection and quantification of PCR products. The qPCR run was performed with primer set 50509/50510 binding neomycin resistance cassette (*Neo*) within the transgene. A conserved endogenous housekeeping gene, cytoglobin b (*Cygb*, primer set AMa09/10) was co-amplified and used as an internal standard for normalizing the amount of DNA used in each different sample. *Neo* and *Cygb* were amplified simultaneously in different tubes and all reactions were run in triplicate. Post-run analysis was done with iCycler iQ Software and the C_t value (defined as the level at which the fluorescence exceeds a set baseline) was determined by using the instrument's software. To determine transgene copy number *Cygb*-standardized transgene C_t values (*Neo*) of samples were compared to reference values of mouse DNA containing exactly one copy of the *Neo* cassette (generously supplied by Dr. Aswin Mangerich, University of Konstanz). The transgene copy number of this mouse was set as 1.0. The relative quantification of transgene copy number with the target sequence *Neo* and the reference sequence *Cygb* (endogenous control) is based on equal PCR efficiencies for both the target (*Neo*) and the internal control (*Cygb*) gene, in order to calculate the initial concentration of the sample on the basis of the C_t value. For this purpose, the PCR efficiency for both reactions was determined by 5-fold serial dilutions of DNA concentrations between 0.8, 4, 20 and 100 ng measured in triplicate. After ensuring that both reactions had the same PCR efficiency, the transgene copy number was estimated according to the $2^{-(\Delta\Delta C_t)}$ method (Livak and Schmittgen, 2001):

$$\text{Transgene copy number} = 2^{-(\Delta\Delta C_t)}$$

$$\Delta C_t = C_t \text{ (standard, } Neo\text{)} - C_t \text{ (endogenous control, } Cygb\text{)}$$

$$\Delta\Delta C_t = \Delta C_t \text{ (unknown sample)} - \Delta C_t \text{ (reference sample)}$$

In addition, a melt curve analysis of all samples was performed to ensure that the PCR had run with high specificity.

3.2.6 Cell culture

3.2.6.1 Cell number determination

Cells were counted either with Neubauer chamber or with Casy Model TT (Schärfe Systems). For each individual cell line, the counting parameters were estimated according to the manufacturer's instructions. Prior to counting with the Casy Model TT, the cell suspension was diluted 1:400 in 10 ml CasyTon®. For each measurement 400 µl were necessary, which was repeated three times.

3.2.6.2 Thawing of cells

Each cell line was used for a maximum of 15 passages. Fresh aliquots of cells frozen in liquid nitrogen and kept in a cryovial, were transferred to a 37°C water bath and kept there until thawing. Thereafter, the cells were transferred into a 50 ml falcon tube and 1 ml of pre-warmed DMEM medium (37°C) was added dropwise under constant shaking. Then a stepwise addition of 2, 4 and 8 ml of pre-warmed DMEM medium in time intervals of 1 min was performed. Cells were then centrifuged for 5 min at 200 g at RT. The cell pellet was then resuspended in fresh pre-warmed DMEM medium and incubated according to the manufacturer's protocol.

Material and methods

3.2.6.3 Cryopreservation of cells

For their long term storage, cells were cryopreserved in 2 ml cryovials in liquid nitrogen. For this purpose, cells were first counted and then adjusted to 1×10^7 cells/ml in freezing medium. Thereafter, 1 ml aliquots were transferred into a pre-cooled (4°C) cryovial and immediately placed into a pre-cooled (4°C) cryo freezer (cooling down 1°C/min; Mr. Frosty) and subsequently placed in a -70°C freezer. Finally, after 4 h the cryovials were transferred from -70°C into a liquid nitrogen storage container.

3.2.6.4 Cell culture and passaging of cells

El-4 suspension cells were passaged every 1 to 3 days and were grown in a range of 1×10^5 - 1×10^6 cells/ml. Cells were aspirated in El-4 cell medium and transferred into a 50 ml falcon tube, centrifuged at 200 g for 5 min at RT. Thereafter, the supernatant was discarded and the remaining cell pellet dissolved in 10 ml pre-warmed El-4 cell medium. Cells were counted, resuspended in an appropriate volume of El-4 cell medium and transferred into a cell culture flask (75 cm^2) at 37°C and 5% CO_2. All adherent cell lines (CO60, COR4, COMF10, HEK293T, 3T3) were passaged every 1 to 3 days, until a desired confluence was reached. After the supernatant was discarded, cells were washed with PBS and then incubated with trypsin/EDTA (0.25%/1 mM) for 1 to 3 min, until cells were visibly detached from the flask surface. Thereafter, trypsination was stopped by resuspending the cells in appropriate cell culture medium with 10% FCS (10 ml) and transferred into a 50 ml falcon tube. Following steps were identical as described above for the EL-4 suspension cells.

3.2.6.5 Transient transfection of EL-4 cells with jetPEI™

For transient transfection of EL-4 cells with the hPARP-1 expression plasmid pwpt-*hPARP-1*, the jetPEI™ solution was used according to the manufacturer's instructions. 2×10^5 Cells/ml were treated with 1 µg DNA and 1.2 µl jetPEI™ solution for 48 h and then used for DNA repair measurements with the FADU technique as described in section 3.2.1.1.

3.2.7 Molecular biological methods

3.2.7.1 Restriction analysis

Restriction digest incubation experiments were done overnight in a total volume of 10 - 20 µl, consisting of 0.1 - 5 µg DNA and 1 - 2 µl of the corresponding restriction enzyme (10 - 20 U) according to the manufacturer's instructions.

3.2.7.2 Agarose gel electrophoresis

To determine the size and quantity of DNA fragments after a restriction digestion or a PCR reaction, DNA was mixed with loading buffer and loaded onto an agarose gel together with a molecular size ladder, separated by agarose gel electrophoresis and stained with ethidium bromide. Depending on the length of the DNA fragment, the percentage of agarose dissolved in TAE electrophoresis buffer varied from 0.3% (w/v) for large fragments (10 kbp) up to 3% (w/v) for very small fragments (100 bp). The gel electrophoresis was performed at a voltage of 120 V in TAE electrophoresis buffer. Thereafter, the gel was stained in ethidium bromide solution (1 µg/ml) for 10 min, followed by 10 min incubation in water in order to wash out the dye from the gel, which was not intercalated into DNA. The DNA in the gel was then visualized under UV light using an Intas gel documentation device.

3.2.7.3 Gel extraction

Gel extraction was performed with the Qiaquick gel extraction kit according to the manufacturer's instructions. In case of small amounts of DNA material, the designated gel slice was layered on the surface of a 0.5 ml reaction tube half-filled with glass wool supplied with a small hole on its tip (custom-made). After centrifugation (900 g, 2 min) the DNA containing eluate was purified by phenol-chloroform extraction and finally dissolved in H_2O.

3.2.7.4 Phenol-chloroform extraction of DNA

An extraction by phenol-chloroform was used to remove proteins from DNA samples. An equal volume of phenol was added to the DNA mixture, vortexed for 10 s and centrifuged at 13,000 g for 1 min. From the resulting three phases, the upper aqueous phase was carefully transferred into a new reaction tube and an equal volume of chloroform/isoamyl alcohol (24:1, v/v) was added to remove phenol, vortexed for 10 s and centrifuged at 13,000 g for 1 min. Thereafter, the aqueous phase was transferred into a new reaction tube and 0.1 volumes of 3 M sodium acetate (pH 5.2) and 3 volumes of 100% ethanol were added, which caused precipitation of DNA within 1 h at -70°C or overnight at -20°C. The samples were then centrifuged at 13,000 g for 30 min at 4°C, the supernatant was discarded and the resulting pellet washed once in 1 ml of chilled 75% ethanol. After the samples had been again centrifuged (13,000 g, 10 min, 4°C), the supernatant was discarded and the resulting pellet air-dried and subsequently dissolved in a volume of 50 to 100 µl MilliQ H_2O.

3.2.7.5 Blunting and dephosphorylation of DNA fragments

For generation of DNA blunt ends within the transgene vector, T4 DNA polymerase was used according to the manufacturer's instructions. In order to prevent a possible religation of the blunted DNA fragments, 5´phophate groups were removed with Antarctic phosphatase according to the protocol of the manufacturer.

3.2.7.6 Purification of reaction mixtures

For the purification of DNA, reaction mixtures were subjected either to a phenol-chloroform extraction or to the procedure of "MiniElute reaction cleanup kit" according to the manufacturer's instructions followed by elution in H_2O.

3.2.7.7 Ligation

The molar ratio necessary to perform a successful ligation reaction between vector and insert was calculated to be 1:5 according to the formula:

$$$$

(bp = length of DNA in base pairs)

Ligation was performed either with T4 DNA Ligase or with Ready-To-Go^{TM} ligase kit at 16°C overnight according to the manufacturer's instructions. The mixture was then used for transfection of *E. coli* DH5α (see section 3.2.9).

3.2.8 Preparation of chemo-competent *E. coli* DH5α

A volume of 4 ml *E. coli* DH5α grown overnight in 2YT medium was used to inoculate 400 ml of the same medium. After reaching a cell density of $OD_{595\ nm}$= 0.6, cells were chilled on ice, centrifuged at 3000 g for 15 min at 4°C and resuspended in 100 ml cold TB buffer. After 10 min, the cells were centrifuged (3000 g, 15 min, 4°C) and resuspended in 30 ml cold TB buffer. Finally, after addition of 7% DMSO and further 10 min incubation, cells were aliquoted into 100 µl portions, snap frozen in liquid nitrogen and stored at -80°C.

3.2.9 Transformation

Samples of 100 µl of chemo-competent *E. coli* DH5α aliquots were thawed on ice and immediately gently mixed with 1 µl of the ligation reaction mixture defined above. Cells were kept on ice for 30 min, then subjected to heat shock at 42°C for 25 s and further incubated for 2 min on ice. Thereafter, cells were mixed with 950 µl LB or SOC medium without antibiotics and incubated at 37°C for 1 h on a thermomixer at 1000 rpm. Finally, the cells were centrifugated at 3000 g for 5 min, the resulting pellet was resuspended in 100 µl LB medium and plated on LB-amp agar plates containing antibiotic (ampicillin, 100 µg/ml) for positive selection of the desired clones. The cells were incubated overnight at 37°C and the resulting colonies were screened for correct vector incorporation on the next day (see section below).

3.2.10 Colony screening

A number of 10 - 20 colonies growing on LB-amp agar plates were picked with a sterile toothpick, then transferred into 3 ml of LB-amp medium (ampicillin, 100 µg/ml) and incubated at 37°C for 12 to 16 h at 250 rpm. After centrifugation of the transformed *E. coli* DH5α cells (3000 g, 5 min, 4°C), DNA was isolated with QIAprep® spin miniprep kit according to the manufacturer's instructions. The successful incorporation of the vector/insert product was checked by restriction digest and agarose gel analysis.

3.2.11 Generation of *hPARP-1* transgenic mice by DNA microinjection

3.2.11.1 Preparation of the transgene for DNA microinjection

The transgene construct pucTE5 obtained after several cloning steps was linearized overnight with restriction enzyme *Pfl*23II according to the manufacturer's protocol. To ensure an efficient digestion and purification of the transgene pUCTE5, it was loaded on two pockets of an agarose gel (0.5%). After electrophoresis the gel was cut vertically between the two pockets. To prevent DNA damage by ethidium bromide and UV light, only one gel half was stained with ethidium bromide and the correct linearized fragment was then marked under UV light. The corresponding intact fragment located on the same position as on the other gel half, was cut out, purified with the Qiaquick gel extraction kit according to the manufacturer's instructions and dissolved in 10 mM Tris-HCl pH 8.0. This extract was sterilized through a Millipore Millex-GV 0.22 µm filter by the use of a 1 ml syringe. Thereafter, the DNA was dialyzed by carefully pipetting it on a Millipore membrane VMW PO2500/VM 0.05 µm floating on 100 ml microinjection buffer at 4°C for 3 h. The DNA concentration was determined and adjusted to 10 ng/µl by photometric measurement and agarose gel electrophoresis, including DNA mass ruler marker of known size and concentration. DNA was stored at 4°C until its microinjection into one of the pronuclei of mouse zygotes.

3.2.11.2 Embryo isolation from mice

To increase the oocyte production in mice, donor females were superovulated by i.p. injection of gonadotropin and followed by chorionic gonadotropin 47 h later. Thereafter, donor females were mated with fertile males and successful mating was identified by formation of vaginal plugs. After cervical dislocation of those mice, zygotes were taken and prepared under the microscope (Figure 6) prior to injection of DNA.

3.2.11.3 DNA injection

After fixing the zygote by means of a glass holding pipette at negative pressure, 1 - 2 pl transgene DNA solution (pUCTE5; 1 - 3 ng/µl) was injected in one of the pronuclei of the 1-cell zygote (Figure 6). For recovery, the zygotes were kept then in a CO_2 incubator at 37°C for 2 h.

3.2.11.4 Embryo transfer in foster females

Eight to ten zygotes were transferred via glass capillaries into the infundibulum of the oviduct of a foster female (Figure 6), which had previously been mated with a vasectomized male. The pups born three weeks later by these foster mice were then analyzed for successful transgene insertion by PCR (see section 3.2.4).

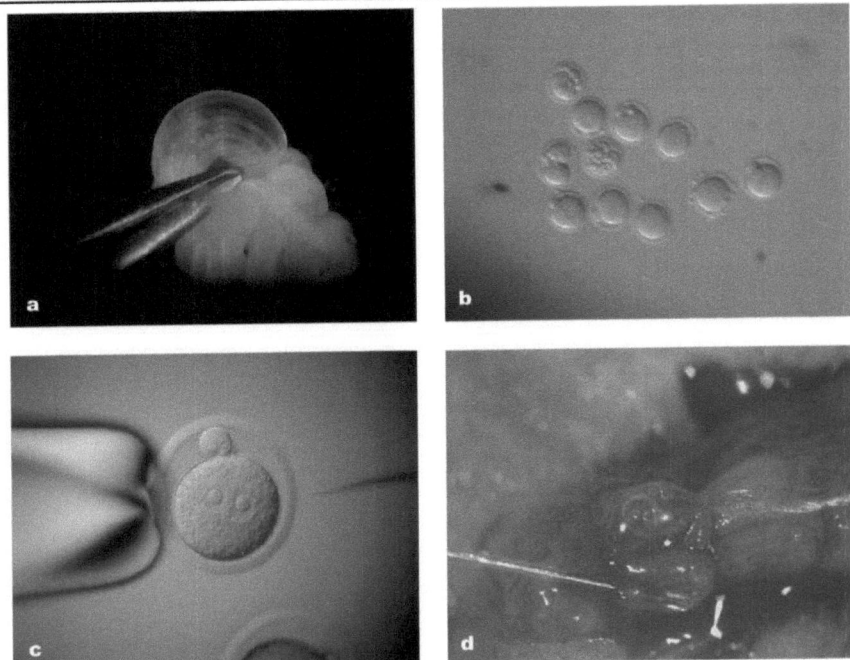

Figure 6: Generation of transgenic mice. (a) Zygotes were harvested 0.5 days post coitum by cervical dislocation. (b) Harvested zygotes. (c) DNA microinjection with a glass needle into the pronucleus of a fertilized ovum. (d) Uterine transfer of the transgene containing zygote. (Photographs were generously supplied by U. Kloz and F. van der Hoeven, German Cancer Research Institute (DKFZ), Heidelberg, Germany).

3.2.12 Protein characterization by SDS-PAGE

Proteins were separated and analyzed according to their molecular mass by denaturating sodium dodecyl sulfate (SDS) polyacrylamide gel electrophoresis (PAGE) (Laemmli, 1970) (8% or 10%; see Table 16). SDS-PAGE was performed in a Hoefer miniVE vertical electrophoresis system (Amersham Biosciences) according to the manufacturer's instructions. Protein samples were dissolved in 1.5x high-urea protein loading buffer and 1x Laemmli buffer (2:1, v/v) and heated to 95°C for 5 min before loading onto the gel. The marker, consisting of 1.5 µl biotinylated wide range molecular weight marker in 13.5 µl 1x Laemmli buffer and 25 µl 2x sample buffer, was then heated to 95°C for 5 min. After cooling to RT, 10 µl of prestained molecular weight marker was added and the complete mixture loaded onto the gel. Samples were run at 10 mA in the stacking gel and at 20 mA in the separating gel.

Table 16: SDS-PAGE separating/stacking gel

SDS-PAGE separating gel	8%	10%
Milli-Q H$_2$O	8.04 ml	7.04 ml
Resolving gel buffer	3.2 ml	3.2 ml
30% Acrylamide/bisacrylamide (37.5:1)	4.2 ml	5.2 ml
10% SDS (w/v)	158 µl	158 µl
10% APS (w/v)	132 µl	132 µl
TEMED	32 µl	32 µl
SDS-PAGE stacking gel	**3%**	
Milli-Q H$_2$O	3.75 ml	
Stacking gel buffer	0.62 ml	
30% Acrylamide/bisacrylamide (37.5:1)	0.5 ml	
10% SDS (w/v)	50 µl	
10% APS (w/v)	75 µl	
TEMED	5 µl	

3.2.13 Western blot

Wet blot was run with a nitrocellulose membrane (Hybond-ECL) in ice cold towin buffer for 2 h at 300 mA. Afterwards, membrane was washed in TNT buffer for 5 min and incubated in blocking solution for 1 h. After cutting off the marker lane from the membrane, the residual membrane was incubated for 1 h with the first antibody (diluted in an adequate concentration in blocking solution). Following three washing processes for 10 min in TNT, the membrane was incubated with a second antibody (HRP-conjugated) in blocking solution, while the biotinylated marker membrane was incubated in streptavidin-conjugated antibody (1:5000) in TNT buffer for 1 h. The membrane was then again washed three times in TNT, incubated for 1 min in ECL solution (mixture of solution A + solution B, ratio 1:1) and chemiluminescence was detected after appropriate exposure times from 5 s up to 15 min with a CCD camera using a FujiLAS 1000 imaging station.

3.2.14 Immunofluorescence

3.2.14.1 Analysis of *hPARP-1* from adherent cells

All adherent cells were seeded on sterile glass coverslips (12 mm) at a density of 2×10^4 cells/cm^2 in a 12-well culture dish and were allowed to adhere overnight. Cells were then washed once with

PBS (2 ml) and fixed with 4% paraformaldehyde (1 ml) for 20 min at RT. Subsequently, the supernatant was removed and cells were incubated with glycine solution for 2 min. After supernatant removal, cells were permeabilised for 5 min with 0.4% triton X-100 in PBS following three washings with PBS. Cells were then incubated in a humid chamber (custom made) with monoclonal antibody FI-23 directed against *hPARP-1* (dilution of 1:250 in blocking solution) at 37°C for 1 h. After three washings with PBS, antibody-antigen complexes were detected with Alexa Fluor488-conjugated goat anti-mouse secondary antibody at 37°C for 45 min. The cells were washed three times, counterstained with Hoechst 33342, washed four times, mounted on a glass slide and were finally examined under a fluorescence microscope for detection of hPARP-1 protein.

3.2.14.2 Analysis of PAR in suspension cells

To analyze PAR formation in EL-4 cells after X-irradiation, cells were grown to 5×10^5 cells/ml, centrifuged at 200 g for 5 min and were then resuspended and counted. Cell number was adjusted to 1×10^6 cells/ml and aliquoted to 100 µl/Eppendorf tube (2 ml). After addition of PARP inhibitors (PJ34, BYK204165 or BYK236864), cells were incubated at 37°C for 30 min. Cells were then X-irradiated on ice at 40 Gy. After that PAR formation could take place for 3 min in a water bath at 37°C. Thereafter, cells were kept on ice to prevent PAR degradation by PARG. Cells were then transferred quickly in the cone-shaped borehole of a cytospin container and centrifuged at 200 g for 5 min at 0°C to stick them to the glass slide underneath. Following this, the cells were then air dried with nitrogen gas and completely dried by exposing them at 37°C for 15 min. Cells were then fixed with pre-cooled (-20°C) methanol/acetic acid (3:1, v/v) for 10 min on ice. The following steps were performed in the same way as in section "analysis of PAR in fibroblasts".

3.2.14.3 Analysis of PAR in fibroblasts

3T3 cells from *Parp-1$^{+/+}$* and *Parp-1$^{-/-}$* mice were grown and passaged in 3T3 cell culture medium to confluence. Then, the cells were washed in PBS, trypsinised and plated on sterile coverslips (12 mm) at a density of 2×10^4 cells/cm^2 in 12-well culture dishes, and were allowed to adhere overnight (Wagner *et al.*, 2007). After exposure of the cultures to PARP inhibitors (BYK204165 or BYK236864; 0.3 – 10 µM, final DMSO concentration 0.3%) for 30 min, cells were first washed with PBS and after that PAR formation was stimulated by treatment with H_2O_2 (5 mM for *Parp-1$^{+/+}$* fibroblasts, 50 mM for *Parp-1$^{-/-}$* fibroblasts) for 5 min at 37°C. The cells were then fixed with methanol/acetic acid (3:1, v/v) for 10 min at RT. After three washings with PBS, cells were incubated with monoclonal antibody 10H directed against PAR at a dilution of 1:250 in blocking solution for 1 h at 37°C in a humid chamber. After three washings with PBS, antibody-antigen complexes were detected with Alexa Fluor488-conjugated goat anti-mouse secondary antibody for 45 min at 37°C. The cells were washed three times and counterstained with DAPI. Coverslips were mounted on glass slides and examined under a fluorescence microscope for detection of PAR.

3.2.14.4 Cytotoxicity assay (microscopy)

After DNA damage in COR4 and COMF10 cells by alkylating agents (MNNG, MMS), applied in the absence or presence of dexamethasone (Dex), apoptosis, necrosis and cell viability were measured by microscopy. COR4 cells (Meyer *et al.*, 2000) are stable transfectants expressing human glucocorticoid receptor derived from the parental SV40 transformed Chinese ovary hamster cells (Kupper *et al.*, 1995) and served as a control cell line. COMF10 cells have in addition to COR4 cell line a plasmid (pPARP93), encoding the full length human *PARP-1* cDNA under the transcriptional control of the mouse mammary tumor virus long terminal repeat (*MMTV-LTR*) promoter, and *pTKneo* (Meyer *et al.*, 2000). The *MMTV* promoter is induced by glucocorticoid hormone (here: Dex) via the glucocorticoid receptor, which then leads to hPARP-1 expression in the COMF10 cells, but not in the COR4 cells, due to the missing pPARP93 plasmid. Cells were grown for two days in a cell culture flask to 80% confluence without selection antibiotics, washed in PBS, trypsinised, centrifuged (200 g, 5 min, RT) and finally resuspended in a small volume of DMEM medium. Cells were counted and adjusted to 2×10^4 cells/ml in stimulation medium containing Dex (100 nM) or not. Then 100 µl/well (2000 cells/well) were transferred into a flat bottom 96-well plate. After 24 h cells were treated with N-methyl-N'-nitro-N-nitrosoguanidine (MNNG, 2.5 - 20 µM, in triplicate) or methyl methanesulfonate (MMS, 100 - 750 µM, in triplicate) diluted in the corresponding medium (+/- Dex), and were incubated at 37°C for 24 h in a CO_2 incubator. Each well was then treated with 5 µl pre-diluted (1:10 in PBS) Sytox®/Hoechst mix for 5 min. The number of necrotic (Sytox® stained), apoptotic (*i.e.* condensed/fragmented nuclei, Hoechst stained) and healthy cells (*i.e.* round nuclei, Hoechst stained) were scored with a fluorescence microscope. A total number of 1,800 cells were assessed for each experimental condition and cell type by examining 200 cells/well in three independent experiments performed in triplicate.

3.2.15 Cytotoxicity assay (FACS)

A number of 4×10^5 COR4 and COMF10 cells/well were seeded in a 6-well plate in a volume of 3 ml/well without selection antibiotics. After 24 h the supernatant was carefully aspirated and 3 ml

Dex stimulation medium was given to the adhered cells. After further 24 h, the cells were treated with MMS (8 – 750 µM) as described in section 3.2.14.4. The supernatant was carefully aspirated 24 h later and collected in 15 ml falcon tubes. The remaining adherent cells in the 6-well plate were rinsed with 1 ml pre-warmed PBS, which was then aspirated and added to the falcon tube. Cells were trypsinised with 300 µl trypsin/EDTA for 2 min until their detachment from the plate surface, aspirated and pipetted in the falcon tube. Following addition of 5 ml medium the mixture was centrifuged (200 g, 5 min, RT), and after removal of the supernatant the cells were resuspended in 1 ml PBS. Cells were then counted in a Neubauer chamber, centrifuged again and after discarding the supernatant, were then resuspended in annexin binding buffer at a cell number of 1×10^6 cells/ml. In each case, 100 µl cell suspension was given to 5 µl annexin V and 10 µl propidium iodide (final concentration 1 µg/ml) and incubated for 15 min in the dark. Then, 400 µl of annexin binding buffer was added, mixed and incubated on ice until flow cytometric measurement was performed. A total of 10,000 events were measured in the FACS device and were further analyzed using the software FlowJo.

4 Results

4.1 Inhibition of PARP-1/-2 in mouse $Parp-1^{+/+}$ and $Parp-1^{-/-}$ fibroblasts

4.1.1 Detection and selective inhibition of PAR formation in $Parp-1^{+/+}$ and $Parp-1^{-/-}$ fibroblasts

For characterization of two new PARP inhibitors, the imidazoquinolinone BYK236864, and the isoquinolindione BYK204165, PAR formation was determined after PARP activation by H_2O_2 treatment in mouse 3T3 fibroblasts. Previously, both compounds had been characterized with respect to inhibitory potency and selectivity on cell-free recombinant hPARP-1 and murine PARP-2 (mPARP-2), as well as in various cellular systems suitable for detection of PARP mediated effects. BYK236864 was additionally evaluated for its inhibitory effect on infarct size caused by coronary artery occlusion and reperfusion in the anesthetized rat (Eltze et al., 2008). As both PARP inhibitors are relatively insoluble in water, with a maximal attainable concentration in saline of 0.02 mM for BYK236864 and 0.009 mM for BYK204165, they were dissolved in 100% DMSO and further diluted in 10% DMSO to the desired test drug concentration. Final DMSO concentrations did not exceed 0.3%, a concentration known to exert no inhibitory effect on PARP-1 activity (Banasik et al., 2004). BYK204165 inhibited cell-free recombinant hPARP-1 and recombinant mPARP-2 with pIC_{50} values of 7.35 and 5.38, respectively, thus displaying 100-fold selectivity for PARP-1, whereas BYK236864 behaved unselective (pIC_{50} 7.81 and 7.55, respectively) (Eltze et al., 2008). These compounds were investigated for inhibition of PARP activity in $Parp-1^{+/+}$ and $Parp-1^{-/-}$ fibroblasts. To determine the selectivity of BYK204165 for inhibition of PAR formation generated by PARP-1, $Parp-1^{+/+}$ cells were exposed to the inhibitor (0.3 – 10 µM), after that PARP-1 and PARP-2 activity was stimulated by H_2O_2 (5 mM) treatment for 5 min. PAR formation was analyzed with PAR specific antibody 10H by immunofluorescence microscopy. Mouse fibroblasts which were not treated with H_2O_2, showed a slight nuclear soft staining, whereas H_2O_2 treatment elicited an intense granular PAR staining in the cell nuclei as a result of activation of both, PARP-1 and PARP-2 (Figure 7). Increasing concentrations of BYK204165 (0.3, 1 and 3 µM) caused a concentration-dependent decrease of PAR signal, but no further effect at 10 µM (Figure 7), i.e. a residual PAR formation is detectable in the presence of BYK204165, even at 10 µM. Unlike BYK204165, BYK236864 (0.3 - 10 µM) effectively decreased the PAR signal in cell nuclei, which was nearly completely abolished at 3 µM and above (Figure 8). As a common phenomenon, the PAR antibody 10H led to a nuclear and soft cytoplasmic staining at 10 µM BYK236864, although PAR formation was totally abrogated. In order to demonstrate that BYK204165 is able to selectively inhibit PARP-1 and does not affect PAR formation by PARP-2 activity at lower concentrations, $Parp-1^{-/-}$ fibroblasts were used. After their treatment with H_2O_2 (50 mM), PAR formation as a result of activation of PARP-2 only, was much weaker than in $Parp-1^{+/+}$ fibroblasts, as expected, and was abrogated in the presence of BYK236864 (0.3 – 3 µM), but remained unaffected by BYK204165 (0.3 – 3 µM) (Figure 9). The latter finding is perfectly compatible with ongoing PARP-2 activity in both cell lines and clearly demonstrates the high selectivity of BYK204165 for PARP-1. (Further results of the experiments have already been published in (Eltze *et al.*, 2008)

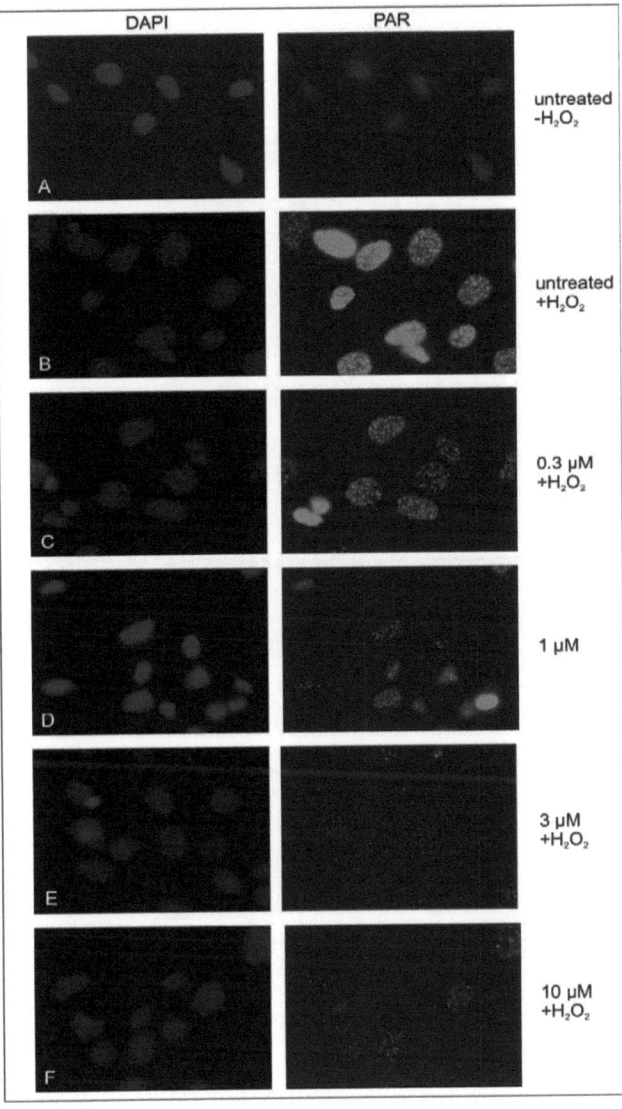

Figure 7: Immunofluorescence analysis of H_2O_2 induced PAR formation in cultured mouse embryonic fibroblasts (3T3) from *Parp-1$^{+/+}$* mice exposed to BYK204165. Cells were treated with 5 mM H_2O_2 and then stained for PAR formation using the primary antibody 10H and the fluorophor-labelled secondary antibody AlexaFluor 488 (FITC channel, right). Nuclei were counterstained by DAPI DNA staining (Hoechst channel, left). No specific PAR staining ("soft" staining) is observed in the absence of H_2O_2 (A). PAR formation induced by H_2O_2 as a result of activation of both PARP-1 and PARP-2 is characterized by a great number of intense, granular signals in the cell nuclei (B). PAR formation decreases with increasing concentrations of BYK204165 (0.3 – 3 µM; C-E), whereas no further decrease in PAR formation is detectable at 10 µM (F).

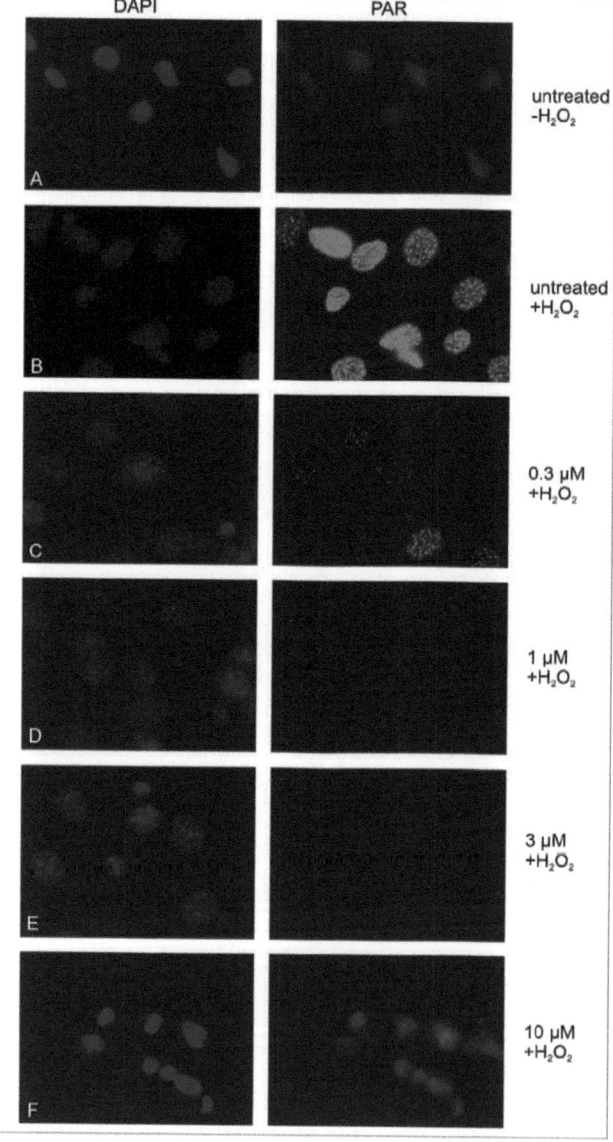

Figure 8: Immunofluorescence analysis of H_2O_2 induced PAR formation in cultured mouse embryonic fibroblasts (3T3) from $Parp-1^{+/+}$ mice exposed to BYK236864. Cells were treated with 5 mM H_2O_2 and then stained for PAR formation using the primary antibody 10H and the fluorophor-labelled secondary antibody AlexaFluor 488 (FITC channel, right). Nuclei were counterstained by DAPI DNA staining (Hoechst channel, left). PAR formation decreases with increasing concentrations of BYK236864 (C, D) and is completely abolished at 3 µM and 10 µM (E, F). Note that the "soft" staining visible at 10 µM is a non-specific, cytoplasmic background (F).

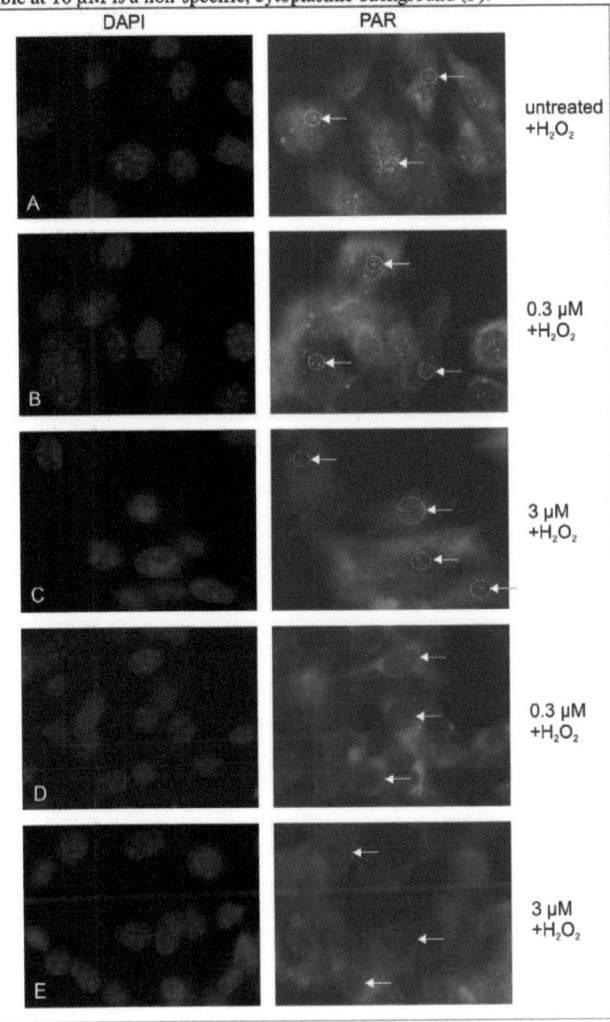

Figure 9: Immunofluorescence analysis of H_2O_2 induced PAR formation in cultured mouse embryonic fibroblasts (3T3) from $Parp-1^{-/-}$ mice exposed to BYK236864 or BYK204165. Cells were treated with 50 mM H_2O_2 and stained for PAR formation using the primary antibody 10H and the fluorophor-labelled secondary antibody AlexaFluor 488 (FITC channel, left). PAR formation (circles) induced by H_2O_2 (untreated, $+H_2O_2$) as a result of activation of PARP-2 only, is abrogated in the presence of BYK236864 (0.3 and 3 µM; D, E), but remains nearly unaffected by BYK204165 (0.3 and 3 µM; B, C). Due to the lack of PAR formation (D, E), the nucleus of the cells appears darker than the

cytoplasm (arrows). Note that a nonspecific, "soft," cytoplasmic background emerges in all *Parp-1*$^{-/-}$ cells that is, however, easily distinguishable from the genuine, granular, intranuclear PAR signals.

4.2 DNA repair and viability in hPARP-1 overexpressing rodent cells

4.2.1 Toxicity induced by alkylating agents in hPARP-1 overexpressing COMF10 cells

PARP-1 has been shown to play a critical role in cell survival. Thus, PARP-1 activation in response to different DNA-damaging stimuli can either lead to necrotic cell death through NAD$^+$ consumption and subsequent energy depletion, or to cellular recovery, depending on the extent of DNA damage stimulus. Since inhibition of PAR formation generally impairs DNA repair mechanisms, the following experiments aimed to explore the effect of dexamethasone (Dex) inducible overexpression of hPARP-1 in Chinese hamster cells (COMF10) on the cytotoxicity induced by the alkylating agents MMS and MNNG. Necrosis, apoptosis and cell viability were measured as experimental endpoints. Additionally, the control cell line COR4 was used, which expresses exogenous glucocorticoid receptor, but no hPARP-1 to exclude the influence of Dex in the assays. As previously shown, the overexpression of hPARP-1 led to about 5-fold increase of overall PARP-1 protein present after Dex stimulation in COMF10 cells compared to Dex untreated cells, while in the presence or absence of Dex the amount of PARP-1 protein in COR4 cells remained unchanged (Meyer *et al.*, 2000). Furthermore, it was demonstrated that about 90% of the COMF10 cell nuclei were stained positive for hPARP-1 following Dex induction (Meyer *et al.*, 2000). Treatment of COMF10 cells with MMS (8 – 750 µM) plus Dex (100 nM) increased necrotic cell death significantly at 500 µM MMS and above, compared to COMF10 cells without Dex (Figure 10, A and B). Under the same conditions, a highly significant and greater increase in necrosis was also observed with MNNG at a concentration of 12.5 µM and above in Dex induced human PARP-1 overexpressing cells compared to Dex untreated cells (Figure 10, C and D).

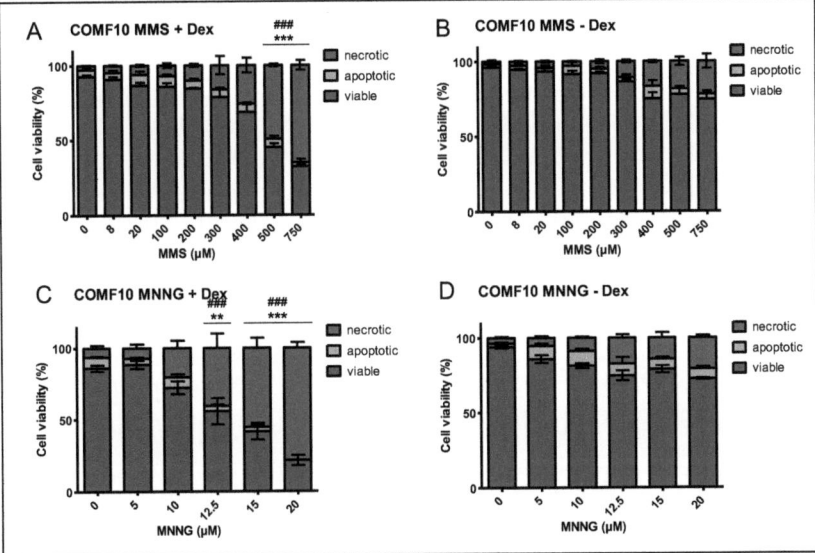

Figure 10: Survival of COMF10 cells. Cells with (A, C) or without (B, D) Dex pretreatment were exposed to MMS or MNNG for 24 h. Viable and apoptotic cells were assessed by Hoechst staining, necrotic cells by SYTOX staining. The percentage of necrotic, apoptotic and viable cells was calculated relative to all Hoechst stained cells. Given are means ± S.E.M. of 3 independent experiments. (**, $p < 0.01$; ***, $p < 0.001$ viable cells, Dex treated compared to Dex untreated; ###, $p < 0.001$ necrotic cells, Dex treated compared to Dex untreated; two-way ANOVA with a Bonferroni posttest).

In order to exclude an effect of a mere expression of the glucocorticoid receptor and treatment with Dex, similar experiments were performed in the control cell line COR4 lacking the hPARP-1 expression plasmid. As expected, Dex treatment had no effect on the parameters viability and necrosis ratio of the cells (Figure 11). However, Dex had a slight, but not significant protective impact on cell survival in MNNG-treated COR4 cells as revealed by a smaller fraction of apoptotic cells compared to Dex untreated COR4 cells (Figure 11, C and D). In this respect, MMS up to 750 µM displayed no protective effect (Figure 11, A and B).

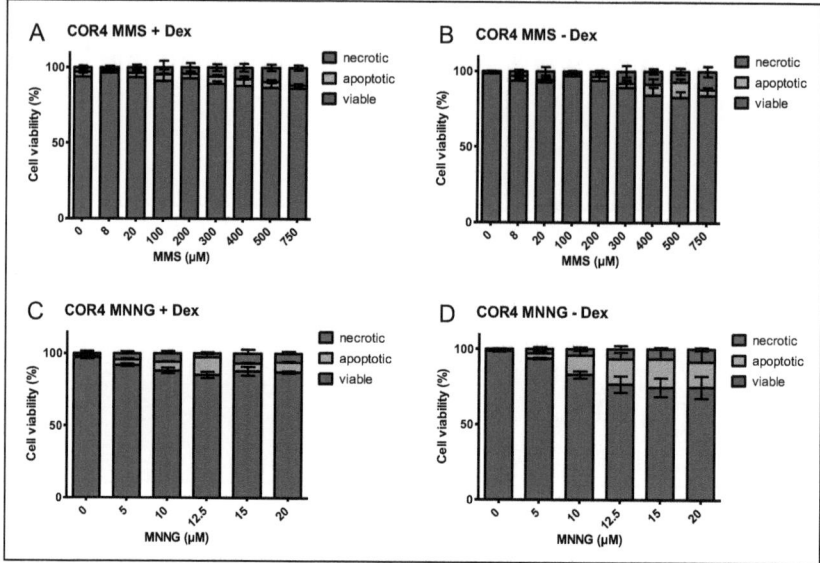

Figure 11: Survival of COR4 cells. COR4 cells with (A, C) or without (B, D) Dex pretreatment were exposed to MMS or MNNG in increasing concentrations for 24 h. Viable and apoptotic cells were assessed by Hoechst staining and necrotic cells by SYTOX staining. The percentage of necrotic, apoptotic and viable cells was calculated relative to all Hoechst stained cells. Given are means ± S.E.M of 3 independent experiments. (No statistical significant differences between groups were observed; two-way ANOVA with a Bonferroni posttest).

4.2.2 Expression of hPARP-1 in murine lymphoma EL-4 cells

In order to determine whether the repair kinetics of DNA damage measured by FADU technique differ between wt and hPARP-1 overexpressing cells, murine lymphoma cells (EL-4) were used in a transient *in vitro* transfection approach to overexpress hPARP-1 protein. EL-4 cells were transfected with pwpt-*hPARP-1* expression plasmid using transfection reagent jetPEI^{TM}, and after 48 h incubation, transfection efficiency was determined under different experimental conditions with a hPARP-1 specific antibody (FI-23) by FACS measurement (Figure 12, A). As a result, the maximal transfection efficiency was determined as 66%, especially obtained with low DNA concentrations (Figure 12, B; Table 17).

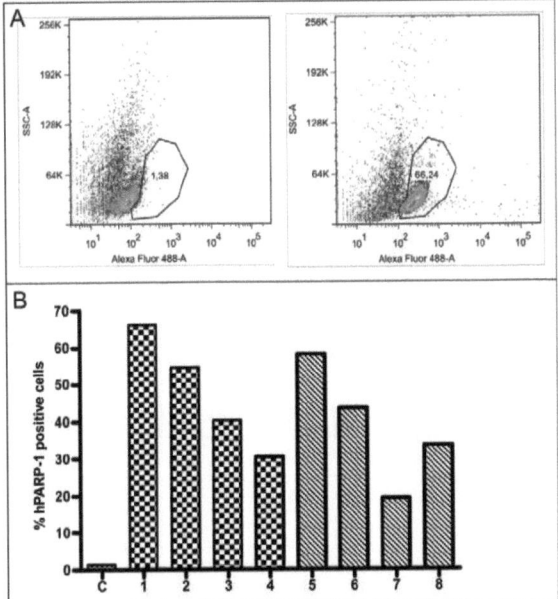

Figure 12: Transfection efficiency of hPARP-1 expression in EL-4 cells. Representative FACS analysis of pwp-*hPARP-1* transfected EL-4 cells labelled with hPARP-1 antibody (FI-23) and conjugated with fluorophor-labelled secondary antibody AlexaFluor 488. (A, left) FACS readout shows control cells which were mock transfected, corresponding to column C in panel B. (A, right) FACS readout represents transfection with pwpt-*hPARP-1*, corresponding to column 1 in panel B. (B) Percentage of hPARP-1 positive cells determined by FACS under different transfection conditions (see Table 17).

Table 17: hPARP-1 transfection efficiency in EL-4 cells under different experimental conditions

Condition	Control	1	2	3	4	5	6	7	8
JetPEI™ reagent [µl]	2	1	1	1	1	2	2	2	2
pwpt-*hPARP-1* DNA [µg]	0	1.2	2.0	3.2	4.0	2.4	4.0	6.4	8
% hPARP-1 positive cells	1.4	66.2	54.6	40.3	30.4	58.1	43.4	19.2	33.4

4.2.3 DNA repair kinetics in EL-4 cells treated with jetPEI™

To exclude that the DNA repair kinetics are influenced by the transfection reagent jetPEI™ itself, EL-4 cells were pretreated with jetPEI™ reagent under the same conditions as pwpt-*hPARP-1* transfected EL-4 cells, and were compared to untreated controls after damage with 7 Gy X-irradiation. Thereafter, DNA repair kinetics were analysed by FADU. As shown in Figure 13, the DNA repair kinetics reached a maximum after 30 min, but were not influenced by jetPEI™ at any time point. (Note: The fluorescence signal was not calculated in gray equivalent, as this experiment was done once only, therefore, it was not possible to calculate S.D. for gray equivalents)

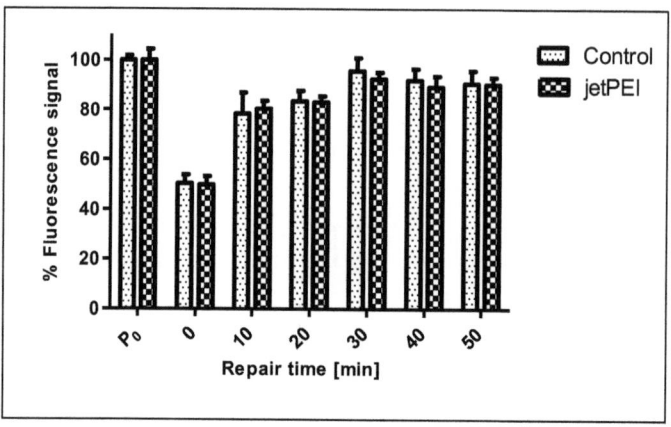

Figure 13: Influence of transfection reagent jetPEI^{TM} on DNA repair. X-irradiation of EL-4 cells with 7 Gy in the presence of transfection reagent (jetPEI^{TM}) or not (control). DNA repair was measured by FADU technique. Experiments were done in quadruplicate and data are expressed as means ± S.D. P_0: Level of fluorescence signal obtained in undamaged cells was set to 100%.

4.2.4 Induction of DNA strand breaks by X-irradiation in EL-4 cells

To determine the reproducibility of the FADU assay in EL-4 cells after their exposure to different X-irradiation doses, cells were exposed to X-irradiation in three independent experiments under identical conditions. As shown in Figure 14, the fluorescence intensity reflecting the amount of intact, double-stranded DNA formation (calculated in % of the maximum level of fluorescence (T, set as 100%), immediately measured after X-irradiation. The fluorescence signal intensity decreases with increasing doses of X-irradiation, indicating a dose-dependent DNA unwinding (linear dose-response relationship up to 8 Gy X-irradiation).

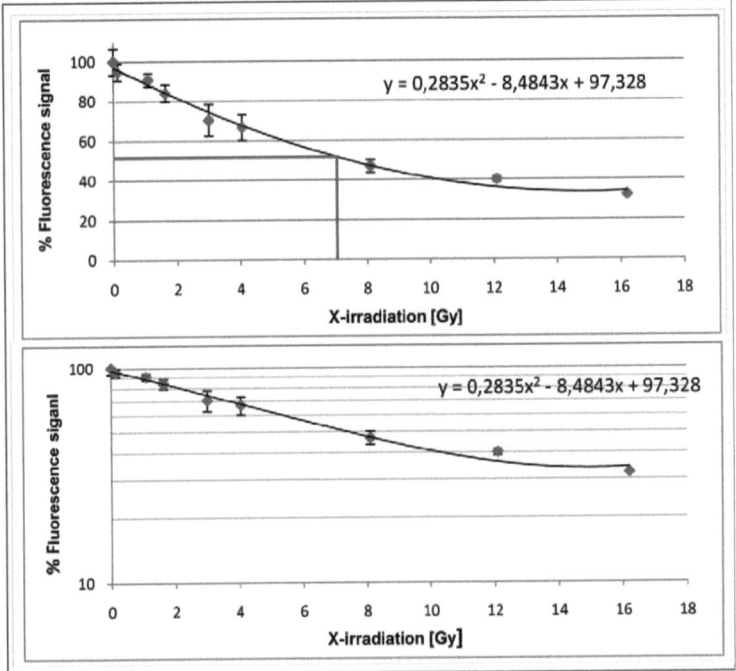

Figure 14: Dose-dependent induction of DNA strand breaks by X-irradiation in EL-4 cells. (Top, linear scale; bottom, logarithmic scale). EL-4 cells were exposed to various X-irradiation doses between 0 and 16 Gy. Given are means ± S.D. from 3 independent experiments, each n = 8. Red line indicates the fluorescence signal at 7 Gy X-irradiation.

4.2.5 Determination of the optimal X-irradiation dose for DNA repair measurements

In order to determine the optimal X-irradiation dose then used for the DNA repair experiments described below, different doses (5, 7 and 9 Gy) of X-irradiation were applied to EL-4 cells, thereafter the cells were incubated at 37°C for various time intervals (0 – 50 min) to allow repair of DNA strand breaks, or were kept on ice. The results of these experiments, in which unwinding of DNA in response to X-irradiation was measured by FADU, are shown in Figure 15. As a result, the higher the X-irradiation dose, the higher the initial DNA damage observed (0 min repair), necessitating increasing time periods to reverse cellular DNA damage to values comparable to unirradiated P_0 controls, i.e. at 5, 7 and 9 Gy these times amounted to 20, 30 and 50 min (± 5%), respectively. Therefore, in the following DNA repair experiments, a dose of 7 Gy was used at which EL-4 cells showed a 50% (± 5%) decrease in fluorescence signal compared to non-irradiated controls (Figure 14, top), thus providing a wide resolution window for detection of DNA repair.

(Note: The fluorescence signal was not calculated in gray equivalent, as this experiment was done once only, therefore it was not possible to calculate S.D. for gray equivalents)

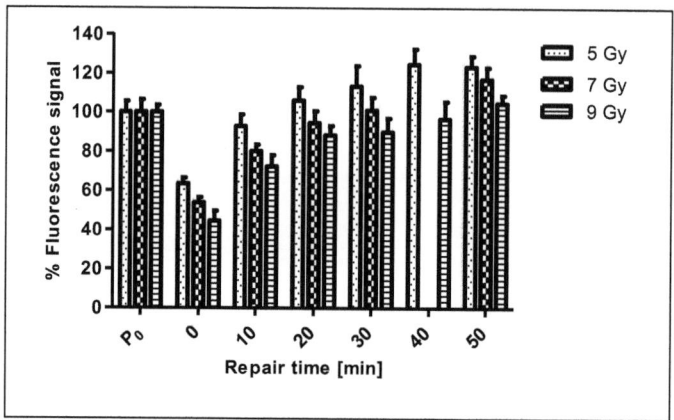

Figure 15: Time course of DNA strand break repair in EL-4 cells. Cells were treated with different doses of X-irradiation and were allowed to repair damaged DNA for different times (0 - 50 min) at 37°C. Given are means ± S.D., n = 8. P_0: Level of fluorescence signal obtained in undamaged cells was set to 100%.

4.2.6 Repair kinetics of DNA strand breaks in EL-4 cells with hPARP-1 overexpression

Repair kinetics of DNA strand breaks induced by X-irradiation were investigated by a transient transfection approach in EL-4 cells. EL-4 cells were transfected with hPARP-1 expression plasmid pwpt-*hPARP-1*, and after the expression of hPARP-1 two days later, the cells were X-irradiated with 7 Gy. DNA strand break repair was measured by FADU at different time periods (0 - 40 min). As shown in Figure 16, at each time point (except 15 min) the level of DNA strand breaks was significantly reduced in hPARP-1 overexpressing cells compared to untransfected cells ($p < 0.05$; two-way ANOVA). However, after 40 min the overall level of repaired DNA in both cell types reached the values of non-irradiated controls (P_0).

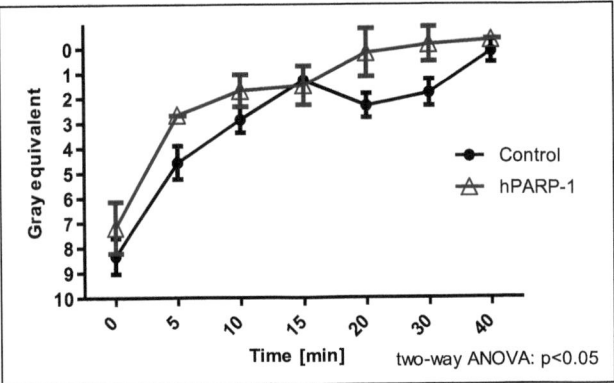

Figure 16: DNA strand break repair in EL-4 cells overexpressing hPARP-1 protein. Cells were X-irradiated with 7 Gy and incubated for different time periods (0 - 40 min) at 37°C for repair of DNA strand breaks measured by FADU technique. Control: EL-4 cells; hPARP-1: EL-4 cells transiently transfected with hPARP-1 expression plasmid (pwpt-*hPARP-1*, transfection efficiency approx. 65%). Given are means ± S.E.M. of three independent experiments each from 4 replicates (control vs. hPARP-1, $p < 0.05$; two-way ANOVA).

4.2.7 Perturbation of PARP activity by PARP inhibition with PJ34 in EL-4 cells

The potent but unselective PARP inhibitor PJ34 (hPARP-1 pK_i 7.72, mPARP-2 pK_i 7.21, (Eltze *et al.*, 2008)) was used for PARP inhibition in EL-4 cells. To determine its inhibitory potency in this cell line, cells were first treated with different PJ34 concentrations (1, 3 and 5 µM) for 40 min followed by X-irradiation with 40 Gy. PAR formation was then detected with PAR specific antibody 10H, conjugated to secondary antibody AlexaFluor 568 and analyzed by immunofluorescence microscopy. After X-irradiation of cells not treated with inhibitor (Figure 17, A), most of the cells showed a clear and high PAR signal intensity, although some cells responded only with a very weak PAR formation. Pretreatment of cells with PJ34 caused a concentration-dependent (1 – 5 µM) and complete inhibition (5 µM) of PAR formation (Figure 17, B-D).

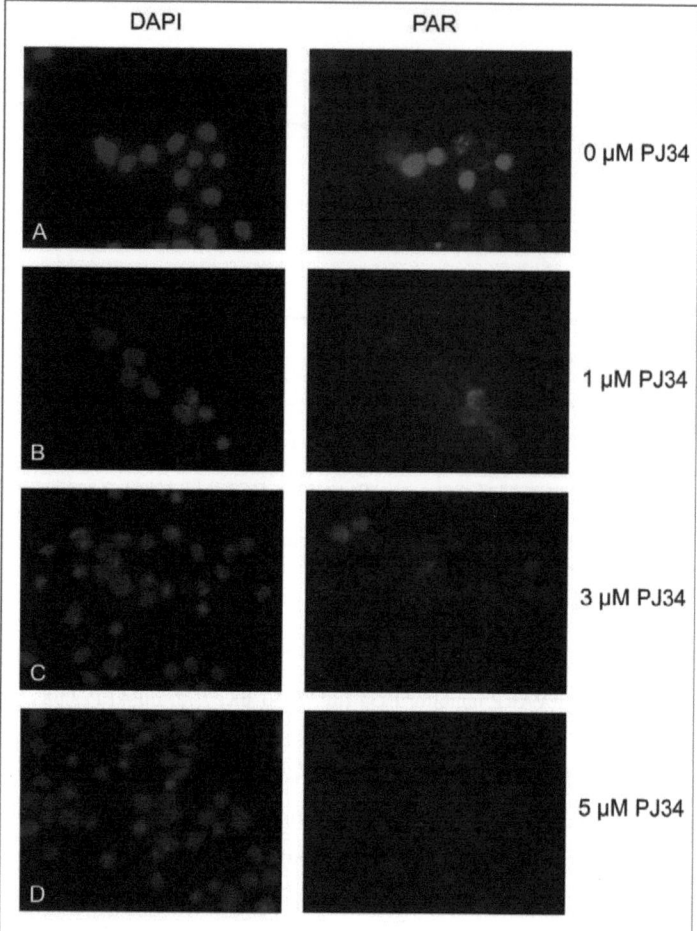

Figure 17: Immunofluorescence analysis of PAR formation induced by X-irradiation in cultured EL-4 cells in the presence of PJ34. Cells were treated with different concentrations PJ34 (1 - 5 µM) for 40 min and X-irradiated with 40 Gy. PAR formation was analyzed using the primary PAR antibody 10H and the fluorophor-labelled secondary antibody AlexaFluor 568 (right), nuclei were counterstained by Hoechst (DAPI) DNA staining (left).

4.2.8 Repair kinetics of DNA strand breaks in EL-4 cells after PARP inhibition

In order to determine the effect of PARP inhibition on DNA repair kinetics in X-irradiated EL-4 cells, PJ34 was used at a concentration of 5 µM, which had been previously shown to completely abolish PAR formation (Figure 18). A 40 min pretreatment of the cells with PJ34 caused a delay in DNA strand break repair at each measured time point, which was significantly different compared to control ($p < 0.01$, two-way ANOVA). By comparing respective time points of DNA repair between control EL-4 cells and those treated with PJ34, a significant difference was detectable at 5 min and 10 min (p < 0.01; Bonferroni posttest). Related to a gray equivalent of 2.5 the delay for repair of DNA strand breaks in EL-4 cells treated with the PARP inhibitor PJ34 amounted to approximately 15 to 20 min compared to untreated cells. Additionally, even after 40 min the PARP inhibited cells (gray equivalent 2.5) were not able to reach the level of repaired strand breaks compared to the control cells (gray equivalent 0.5). Interestingly, the DNA repair kinetics seemed to be similar, as both curves virtually exhibited the same shape.

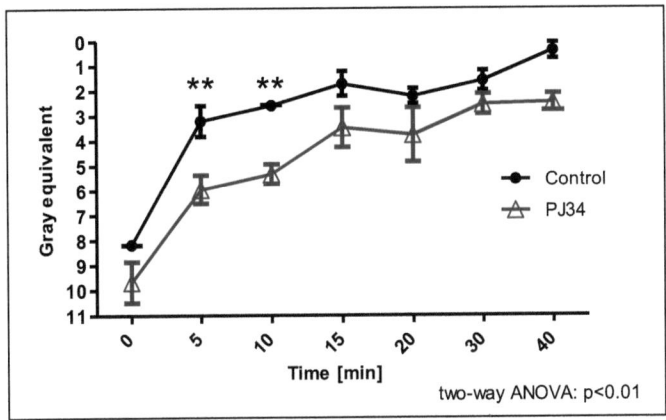

Figure 18: Repair of DNA strand breaks in EL-4 cells after PARP inhibition. EL-4 cells were pretreated with 5 µM PJ34 for 40 min or not, X-irradiated with 7 Gy and incubated for different time periods (0 - 40 min) at 37°C for repair of DNA strand breaks measured by FADU technique. Given are means ± S.E.M. of three independent experiments each from 4 technical replicates. Control curve vs. hPARP-1 curve, $p < 0.01$; two-way ANOVA. **, $p < 0.01$; Bonferroni posttest.

4.3 Experiments for generating *hPARP-1* transgenic mice

4.3.1 Generation of the *hPARP-1* transgene for DNA microinjection

The main objective in this study was to generate transgenic mice with a tissue-specific hPARP-1 protein expression by DNA microinjection of a transgene into mouse zygotes. For this purpose, the classic cloning vector pUC19 was used to generate the transgene containing *hPARP-1* cDNA for DNA microinjection. Since the original multiple cloning site (MCS) of pUC19 was not suitable for the further cloning steps, a completely new MCS was generated and exchanged against the existing one within the pUC19 cloning vector. The new MCS consists of (5´to 3´): *Eco*RI overhang on the 5´end, followed by *Pfl*23II, *Bgl*II, *Xba*I, *Sac*I, *Xho*I, *Sma*I, *Not*I, *Pfl*23II and *Hin*dIII overhang at the 3´end (Figure 19). The MCS was generated by hybridization of the two phosphorylated primers

Results

50303 (sense) and 50304 (antisense). Therefore, the two primers were mixed in an equimolar ratio (10 μM), heated to 95°C and slowly cooled down to 25°C at a rate of 2.5°C/min.

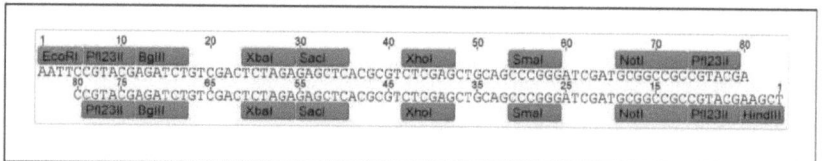

Figure 19: Schematic illustration of the created new MCS for generation of the transgene pUCTE5.

First, to generate the new cloning vector by using pUC19 as a vector backbone, the original MCS from pUC19 vector was excised by *Eco*RI and *Hin*dIII restriction and dephosphorylated by Antarctic phosphatase. Thereafter, the new MCS was inserted by sticky end ligation and transformed into *E. coli* DH5α. Correct insertion was verified with restriction enzymes recognizing specific restriction sites within the new MCS. The generated new cloning vector was designated pUCTE1 (Figure 20, A).

Second, a plasmid named pUCTE2 was generated by insertion of human ubiquitin C promoter (*ubiCp*) into pUCTE1 (Figure 20, B). The promoter was excised from *Ubi*-junB by restriction digest with *Bgl*II and *Xba*I and cloned into pUCTE1 with *Bgl*II and *Xba*I by sticky end ligation and transformation into *E. coli* DH5α. To ensure correct insertion, pUCTE2 was digested with *Bgl*II and *Xba*I.

Third, a further plasmid named pUCTE3 was generated by insertion of the *Neo*/Stop sequence flanked by *loxP* sites, located 3′ from human *ubiCp* (Figure 20, C). To eliminate an existing *Not*I recognition sequence in the PGKneotpAlox2 vector bearing the *Neo*/Stop sequence, this vector was digested with *Not*I. The sticky ends were then removed by Mung bean nuclease digest, followed by blunt end ligation and transformation into *E. coli* DH5α. The correct elimination of the *Not*I recognition site was confirmed by the inability of *Not*I to cause a digest of PGKneotpAlox2, and named PGKneotpAlox2-*Not*I. Then, the floxed *Neo*/Stop sequence was excised with *Sac*I and *Xho*I from PGKneotpAlox2-*Not*I and cloned into pUCTE2 via *Sac*I and *Xho*I ligation.

Fourth, a further plasmid pUCTE4 was generated by insertion of the *hPARP-1* cDNA located 3′from the floxed *Neo*/Stop sequence into pUCTE3 (Figure 20, D). For this purpose, the plasmid pPARP25 was digested with the restriction enzymes *Xho*I and *Sma*I, and the *hPARP-1* cDNA obtained thereof was cloned into pUCTE3 via sticky-/blunt end ligation by *Xho*I and *Sma*I.

Finally, pUCTE5 was generated by cloning the human growth hormone (hGH) minigene into pUCTE4 (Figure 20, E). For this purpose, P1017 was digested with *Bam*HI, then blunted with Mung bean nuclease, digested with *Not*I and afterwards cloned into pUCTE4 via sticky-/blunt end ligation with *Sma*I and *Not*I. The plasmid pUCTE5 was verified for correct sequence by DNA sequencing at GATC (Konstanz, Germany).

Results

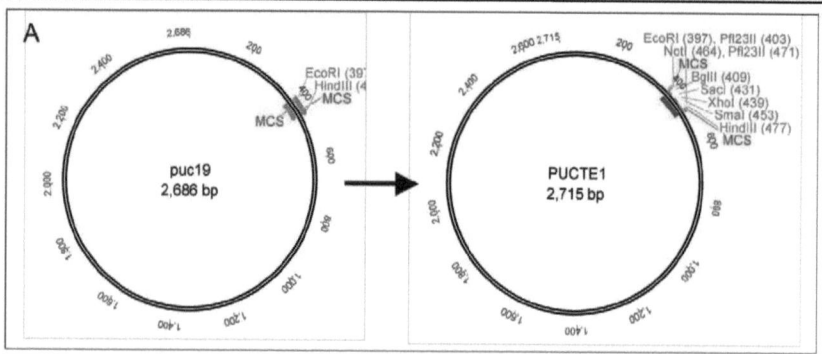

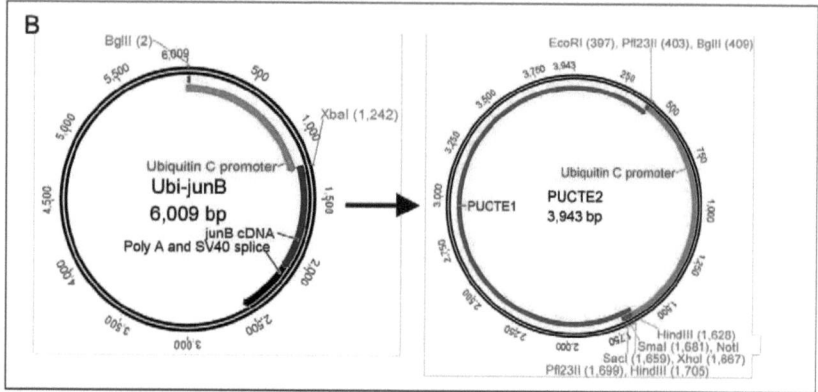

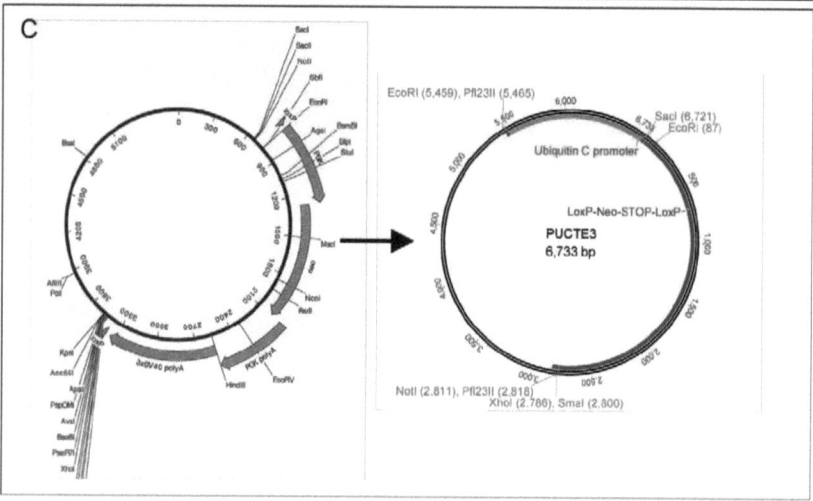

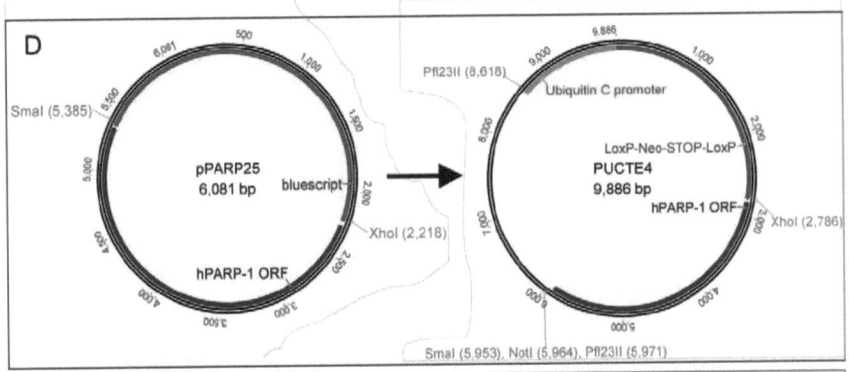

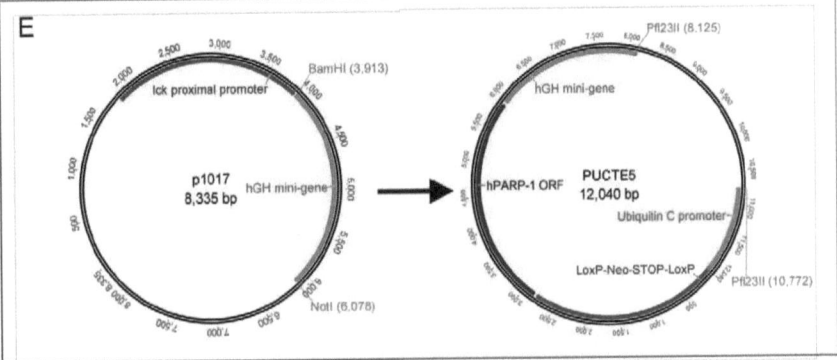

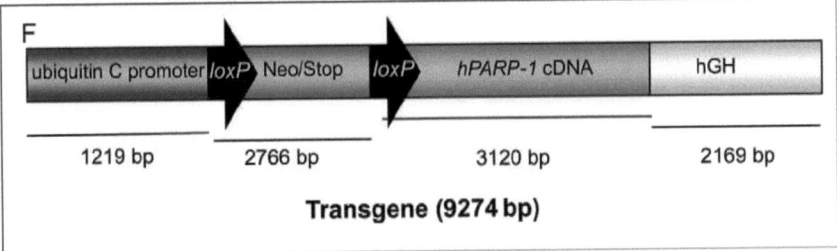

Figure 20: Schematic representation of the plasmids for the transgene generation which was then used for DNA microinjection to generate transgenic mice. (A-E) Plasmids used to generate pUCTE1 to pUCTE5. (F) The transgene consists of: 5′ human ubiquitin C promoter, floxed neomycin resistance cassette with a transcriptional Stop sequence (Neo/Stop), hPARP-1 cDNA and a 3′ untranslated region of human growth hormone (hGH) minigene (exons 1 to 5), bearing a transcription terminator and a polyadenylation sequence for efficient gene expression.

4.3.2 Functional expression analysis of the *hPARP-1* transgene *in vitro*

To assess whether the generated transgene pUCTE5 was able to express hPARP-1 after the excision of the transcriptional *Neo*/Stop sequence, hamster CO60 cells were transfected either with the transgene pUCTE5 alone or in co-transfection with a plasmid bearing C*re* recombinase (turbo-*Cre*).

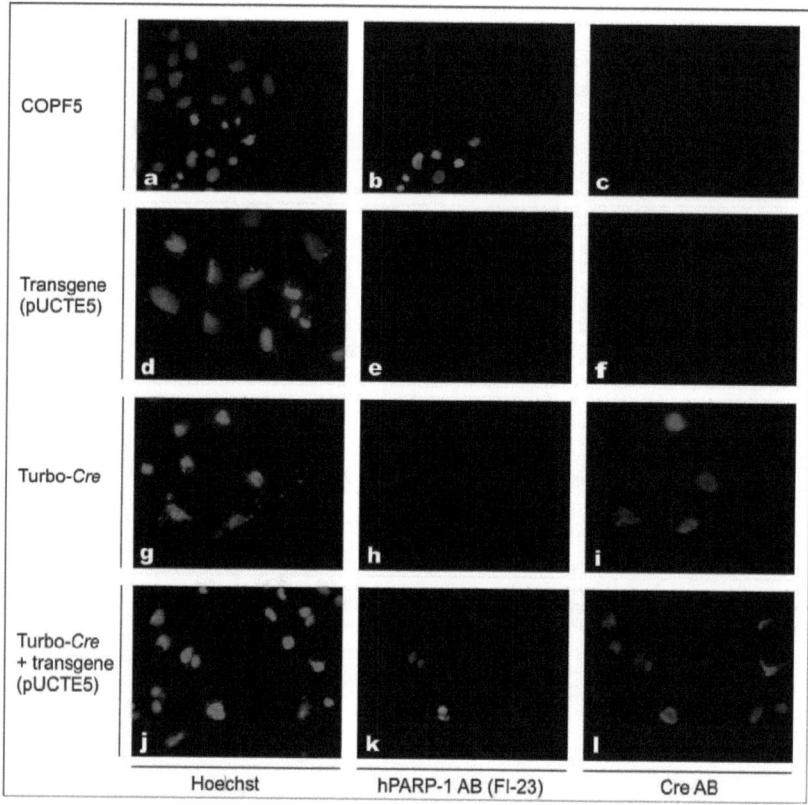

Figure 21: Immunofluorescence staining for detection of transgene function in transfected hamster CO60 cells. Positive control (a-c), transgene pUCTE5 (d-f), turbo-*Cre* (g-i) or both (j-l). Cells transfected only with the expression cassette pUCTE5 (d-f) served as a negative control. Immunofluorescence staining was performed using antibodies (AB) recognizing hPARP-1 (FI-23) (b, e, h, k) and Cre recombinase (Cre) (c, f, i, l). DNA was counterstained using Hoechst (a, d, g, j). Human PARP-1 fluorescence signals were only detected in cells co-transfected with turbo-*Cre* and the transgene pUCTE5 (k), and in stable transfectants expressing hPARP-1 (COPF5) used as a positive control (b).

Expression of the hPARP-1 protein was detected with species-specific anti-hPARP-1 antibody (FI-23) and analyzed by immunofluorescence microscopy. As a positive control, stable transfectants of hPARP-1 expressing cells (COPF5) were used. Cells transfected with either transgene (pUCTE5) or

Cre expression plasmid (turbo-*Cre*) alone did not show any hPARP-1 expression due to the presence of the transcriptional *Neo*/Stop sequence. Cells co-transfected with both plasmids (pUCTE5, turbo-*Cre*) resulted in hPARP-1 expression due to the excision of the *Neo*/Stop sequence by Cre recombinase (Figure 21).

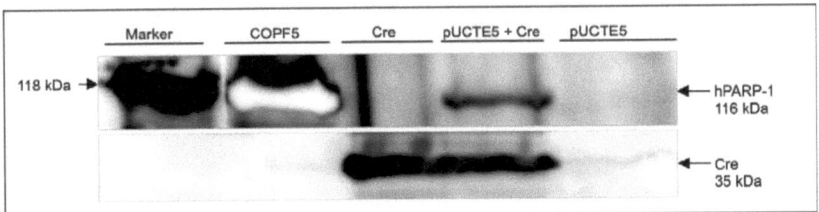

Figure 22: Western blot analysis of hPARP-1 and Cre recombinase in transfected hamster CO60 cells. (Top) Western blot membrane incubated with hPARP-1 specific antibody (FI-23) and HRP-conjugated goat anti-mouse secondary antibody. (Bottom) Blot membrane incubated with Cre antibody (Cre recombinase) and HRP-conjugated goat anti-rabbit secondary antibody. Marker, molecular weight ladder.

After the evidence for successful transfection with pUCTE5 and turbo-*Cre* and expression of hPARP-1 in hamster CO60 cells, the next aim was to determine the extent of protein expression in these cells. Therefore, Western blot analysis was performed with cells transfected in the same conditions as for the immunofluorescence analysis. As shown in Figure 22, a low hPARP-1 protein expression was detectable when Cre recombinase (turbo-*Cre*) was co-transfected with the transgene pUCTE5, but not after transfection with the transgene pUCTE5 alone (Figure 22, top). The Cre recombinase expression was high in cells transfected with turbo-*Cre*, but was not detectable in cells transfected with the transgene pUCTE5 alone (Figure 22, bottom).

4.3.3 Detection of *hPARP-1* transgenic founder mice by real-time PCR

The transgene was excised from pUCTE5 by the restriction enzyme *Pfl*23II, and the resulting linearized transgene obtained (Figure 20, F) was then used for DNA microinjection into mouse zygotes. After the first series of DNA microinjections into mouse zygotes, 79 putative founder mice were born. Three weeks later, DNA was isolated from tail biopsies and analysed by real-time PCR. As shown in Figure 23, only 4 transgenic founder animals out of a number of 79 mice were detected in a real-time PCR analysis using an *hPARP-1* specific probe (microinjection efficiency 5%).

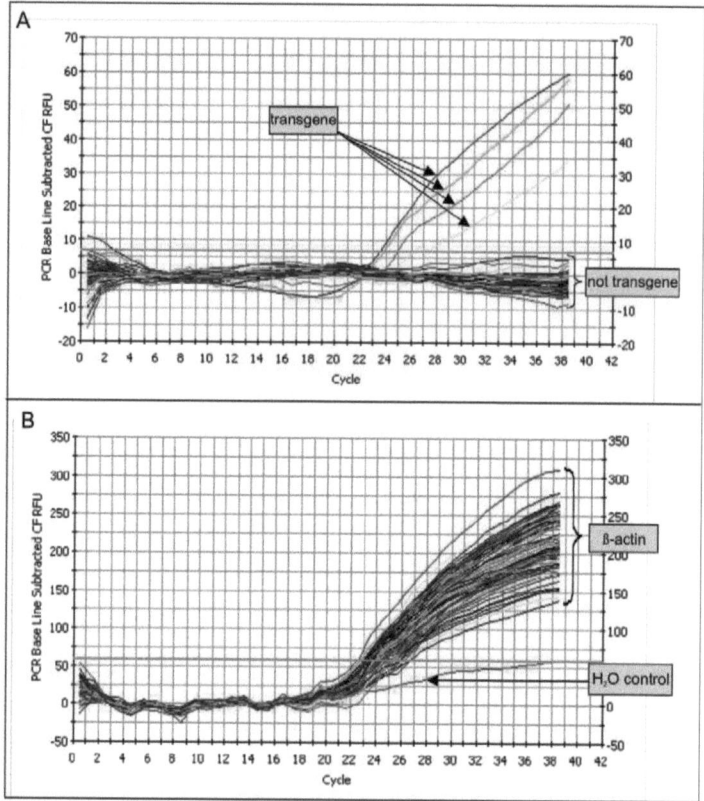

Figure 23: Genotyping of 79 putative hPARP-1 founder mice by real-time PCR analysis. (A) *hPARP-1* detection shown as normalized fluorescence intensities with TaqMan probes during amplification of an *hPARP-1* specific fragment using approximately 100 ng genomic DNA from putative transgenic *hPARP-1* founder mice. (B) *β-Actin* detection shown as normalized fluorescence intensities with TaqMan probes during amplification of a *β-actin* specific fragment using approximately 100 ng genomic DNA from putative transgenic *hPARP-1* founder mice. TaqMan probe was labelled either with HEX (specific binding to *hPARP-1* amplified fragment; primer set: 40804, 40805; probe: 40326) or with FAM (specific binding to murine *β-actin* amplified fragment; primer set: 40327, 40328; probe: 40329). CF RFU, curve fit relative fluorescence units.

To confirm that each PCR well was supplied with sufficient template material of high quality, each well contained an additional primer set (40327, 40328) and a probe (40329) for *β-actin* ("housekeeping gene") as a control. As a result, each well exhibited a significant increase of the fluorescence signal for *β-actin* above the detection threshold, except for the negative control with H_2O.

Results

4.3.4 Detection of *hPARP-1* transgenic founder mice by conventional PCR

In a second series of DNA microinjection into mouse zygotes, only 7 putative transgenic founder mice were born. Therefore, a third attempt of DNA microinjection was made that generated further 61 putative transgenic founder mice. These mice were then analysed after tail biopsies by conventional PCR technique for the presence of the transgene. A representative experiment is depicted in Figure 24. In these experiments, the housekeeping gene *β-actin* was used as a DNA quality control. As a result, the second series of DNA microinjection yielded 3 (microinjection efficiency 42.8%), the third 11 (microinjection efficiency 18%) transgenic founder animals.

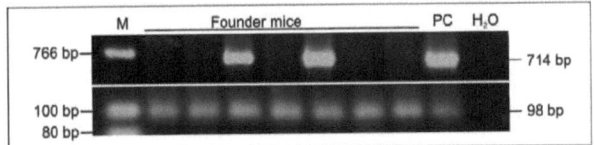

Figure 24: Analysis of founder mice by conventional PCR. Top: *hPARP-1* detection with specific primers binding to the cDNA sequence of *hPARP-1* (expected size: 714 bp; primer set: 30404, 30405). Bottom: *β-Actin* detection with specific primers binding to the murine *β-actin* gene (expected size of 98 bp, primer set: 40237, 40328). M, molecular size ladder; PC, positive control; H$_2$O, water control.

4.3.5 Analysis of hPARP-1 protein level in transgenic mice

For the analysis of hPARP-1 protein, *hPARP-1* transgenic mice from all founder lines obtained were mated with *lck-Cre* transgenic mice, their offspring was analysed for hPARP-1 protein content after isolation of proteins from the thymus. Unexpectedly, it was not possible to detect any hPARP-1 protein in the transgenic mice. A representative Western blot is shown in Figure 25. The protein levels of endogenous mouse PARP-1 detected with CII-10 antibody varied considerably. The extent of Cre protein expression identified by specific antibody (Cre) showed a high variability, ranging from lack of any detection to high expression levels (Figure 25).

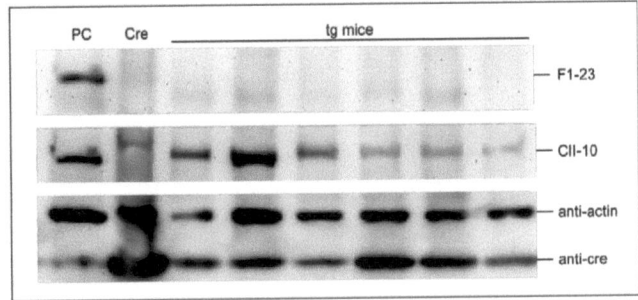

Figure 25: Western blot analysis of hPARP-1 and Cre recombinase expression in *lck-Cre* x *hPARP-1* transgenic mice. Western blot analysis of thymus protein from transgenic mice for hPARP-

1 expression (detection with monoclonal antibody FI-23), general PARP-1 expression (detection with monoclonal antibody CII-10) and Cre recombinase expression (detection with polyclonal Cre antibody). β-Actin as loading control. Cre, CO60 cells transfected with turbo-*Cre* expression plasmid. PC, spleen extracts from an existing hPARP-1 expressing mouse model was used as positive control (Mangerich *et al.*, 2009), obtained from Dr. Aswin Mangerich.

Additionally, *hPARP-1* transgenic mice from all founder lines were mated with *EIIa-Cre* transgenic mice. Their offspring was analyzed for hPARP-1 expression after protein isolation from spleen and kidney (experiments were performed in collaboration with Dr. Aswin Mangerich, University of Konstanz). As shown in a representative Western blot (Figure 26), hPARP-1 protein was not detectable in both organs in any of these transgenic mice, whereas endogenous mouse PARP-1 protein was detectable in the spleen but not in the kidney.

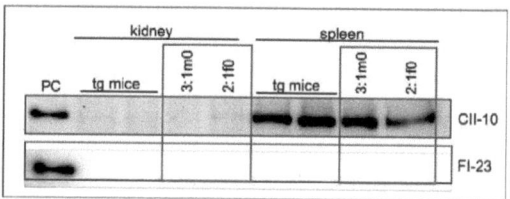

Figure 26: Western blot analysis of PARP-1 expression in kidney and spleen from *hPARP-1* x *EIIa* transgenic mice. Western blot analysis of kidney and spleen from transgenic mice for hPARP-1 expression (detection with monoclonal antibody FI-23) and general PARP-1 expression (detection with monoclonal antibody CII-10). PC, positive control, recombinant hPARP-1 protein, 200 ng. (This experiment was performed in collaboration with Dr. A. Mangerich, University of Konstanz).

4.3.6 Excision of the *Neo*/Stop sequence in genomic DNA of *hPARP-1* transgenic mice

In order to generate transgenic mice expressing hPARP-1 protein, founder animals which have previously been characterized for successful insertion of the hPARP-1 transgene by genotyping, were mated with *lck-Cre* transgenic mice expressing Cre recombinase in T-cells. Their offspring was expected to show an excision of the transcriptional *Neo*/Stop sequence in DNA isolated from thymus. Isolated DNA from thymus of those transgenic mice was used for analysis in a flanking PCR reaction. The primers used (50610, 50611) flanking the *Neo*/Stop sequence, should produce either an amplicon of 3,400 bp in the case of no excision, or an amplicon of 650 bp in the case of successful excision of the *Neo*/Stop sequence (Figure 27, A). As a result, most of the transgenic mice showed a complete excision of the *Neo*/Stop sequence on DNA level (Figure 27, B).

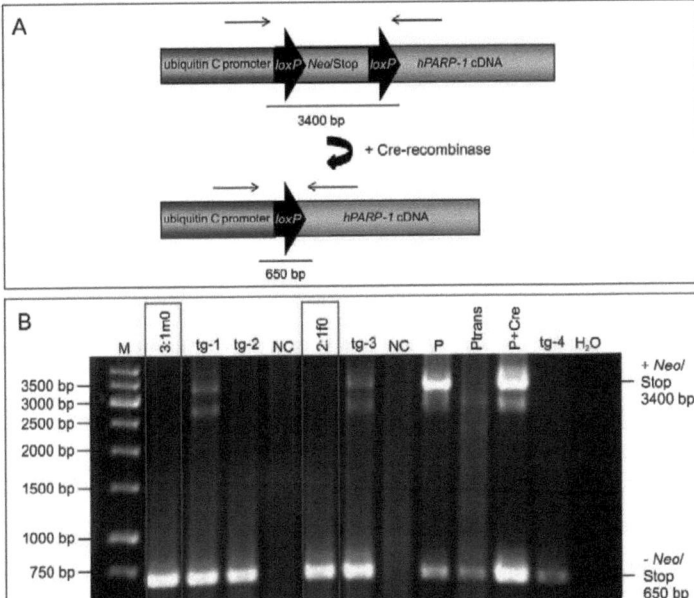

Figure 27: Excision of the Neo/Stop sequence after mating of hPARP-1 x lck-Cre tg mice. (A) Schematic illustration of the excision of the Neo/Stop sequence in vivo by Cre recombinase and the expected length of the amplicons in a PCR reaction before and after successful excision. (B) Flanking PCR from mouse thymus DNA with successful (3:1m0, tg-2, 2:1f0, tg-4) or incomplete excision (tg-1, tg-3) of the Neo/Stop sequence. Controls: NC, negative control with DNA from wt mouse; P, pUCTE5; Ptrans, transfection of pUCTE5 expression cassette in EL-4 cells; P+Cre, co-transfection of pUCTE5 expression plasmid and turbo-Cre expression plasmid. tg, transgenic mouse; H_2O, water control; M, molecular size ladder; m, male; f, female.

As a control, the correct excision of the Neo/Stop sequence could be confirmed in a cellular assay by using EL-4 cells co-transfected with both pUCTE5 and turbo-Cre (Figure 27, B (P + Cre)).

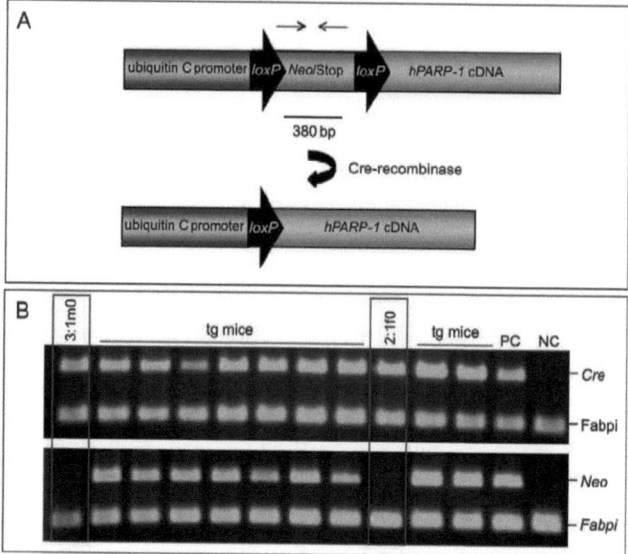

Figure 28: Excision of the *Neo*/Stop sequence after mating of *hPARP-1* x *EIIa-Cre* tg mice. (A) Schematic illustration of the excision of the *Neo*/Stop sequence *in vivo* by Cre recombinase and the expected length of amplicon in a PCR reaction before and after successful excision. (B) PCR with DNA of tail biopsies prepared from different mice. Shown are samples of successful (3:1m0, 2:1f0) or incomplete excision (tg mice) of the *Neo*/Stop sequence. Controls: PC, positive control, *lck-Cre* x *hPARP-1* mouse tail DNA; NC, negative control, wt mouse tail DNA (Experiment was performed in collaboration with Dr. A. Mangerich, University of Konstanz).

In addition, *hPARP-1* transgenic mice were mated with ubiquitously Cre recombinase expressing *EIIa-Cre* mice, the latter mouse type being characterized by expression of Cre recombinase in early embryonic development. This mating should lead to an efficient excision of the *Neo*/Stop sequence in nearly all cell types of the transgenic offspring derived thereof. To prove whether the excision was successful, prepared DNA from tail biopsies was analysed by a slightly modified PCR approach depicted in Figure 28, A (experiments were performed in collaboration with Dr. Aswin Mangerich). The chosen primer set (AMa27/28) bind to the neomycin resistance cassette located between the two *loxP* recognition sites. The PCR reaction should generate an amplicon of 380 bp, if there were cells in which the *Neo*/Stop sequence was not excised. In case of a complete excision no amplicon is synthesised. To ensure that all offspring mice contained the parental *Cre* recombinase gene from *EIIa-Cre* mice, a specific primer set (AMa21/22) for the *Cre* recombinase gene (408 bp) was used.

Furthermore, a control primer set (AMa25/26) for *Fabpi* (200 bp, fatty acid binding protein, intestinal) was used to prove whether the PCR reaction had run properly. As depicted in Figure 28, the same *hPARP-1* transgenic mice (3:1m0, 2:1f0 from Figure 27, B) previously being mated with *lck-Cre* mice and showing an excision of the *Neo*/Stop sequence in the thymus, were also mated with *EIIa-Cre* mice. Their offspring revealed a complete excision of the *Neo*/Stop sequence in DNA from tail biopsies (3:1m0, 2:1f0; Figure 28, B). However, in most of the transgenic mice, excision of the *Neo*/Stop sequence obviously did not occur completely in DNA from tail biopsies, noticeable by a slightly weaker *Neo* amplicon signal compared with control (Figure 28, B).

4.3.7 Analysis of founder mice for complete transgene integration

To test whether the complete sequence of the transgene was integrated into the mouse genome, different primer sets were used, which bind in the promoter-, *Neo*/Stop-, *hPARP-1* cDNA- and hGH minigene-sequence of the inserted pUCTE5 transgene of DNA samples prepared from tail biopsies. As a positive control for the complete sequence, the transgene plasmid pUCTE5 was used for comparison. Figure 29 shows a representative example of 3 to 4 different founder mice. All 18 founder mice from three series of DNA microinjections were tested, and all of them were found to have a complete insertion of the transgene in their genome.

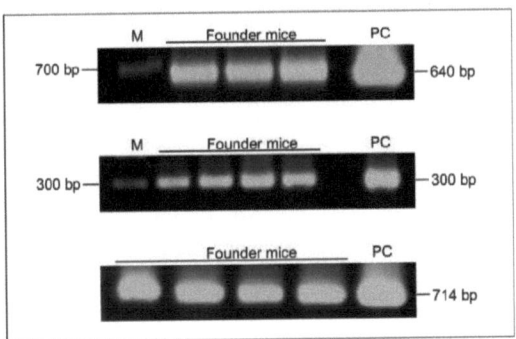

Figure 29: Genotyping of 3 to 4 founder mice in different regions of the transgene. (Top) Primer set (50404, 50405) bind to the *hUbiC* promoter region and to the *Neo*/Stop sequence. Expected size 640 bp. (Middle) Primer set (30302, 50402) bind to the 3´site of *hPARP-1* cDNA and to the 5´site of hGH minigene. Expected size 300 bp. (Bottom) Primer set (30404, 30405) bind to the *hPARP-1* cDNA sequence. Expected size 714 bp. M, molecular size marker; PC, positive control (pUCTE5 plasmid).

4.3.8 Analysis of *hPARP-1* mRNA transcription in transgenic animals

To demonstrate *hPARP-1* mRNA transcription in the *lck-Cre* x *hPARP-1* transgenic mice, a reverse transcriptase PCR (RT-PCR) approach was used that has been shown to be a highly sensitive method to detect mRNA transcripts. RNA was purified from the thymus of transgenic and wt mice, transcribed into complementary DNA (cDNA) by using a retroviral reverse transcriptase together with oligo(dt) primer followed by a PCR reaction which amplified the *hPARP-1* sequence with specific primers (30401/30701). Figure 30 (A) depicts representative experiments from five tested transgenic animals, from which *hPARP-1* mRNA samples transcribed to cDNA and amplified by PCR reaction were loaded onto a 1% agarose gel. In order to prevent false positive results, the RT-PCR was also run without reverse transcriptase prior to the PCR reaction, which should not generate a product in case of DNA-free RNA. As demonstrated in Figure 30, the RNA was free of DNA contamination. To ensure that intact thymus RNA had been isolated and sufficiently transcribed into cDNA, controls with a primer set for murine *β-actin* as the endogenous housekeeping gene were run in parallel (Figure 30, B). As a result, from all 18 transgenic founder lines, only three bore a cDNA transcript of *hPARP-1* mRNA, (2, 4 and 5) but lacked hPARP-1 protein expression (see section 4.3.5).

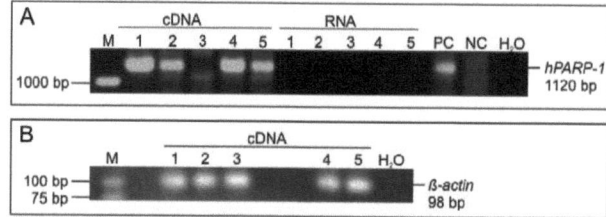

Figure 30: Analysis of mRNA of *hPARP-1* transgenic mice. (A) Analysis of *hPARP-1* mRNA transcription in five transgenic mice with (1-5, cDNA) or without (1-5, RNA) reverse transcriptase in RT-PCR. (B) Analysis of endogenous murine *β-actin* mRNA transcription into cDNA serving as controls. Sample 1 from a mouse tested positive for hPARP-1 expression by Western blot analysis, obtained by Dr. Aswin Mangerich. PC, positive control with DNA from transgenic mouse; NC, negative control with DNA from wt mouse; M, molecular size ladder; H$_2$O, water control.

4.3.9 Relative quantification of the transgene copy number

In order to determine the transgene copy number in all transgenic founder lines, quantitative real-time PCR was used, including primer sets that specifically bind either to the neomycin cassette within the hPARP-1 transgene or to the endogenous gene cytoglobin b (Cygb), which was used as a control to normalize DNA starting quantity. To determine the transgene copy number by the $\Delta\Delta C_t$ method, the amplification efficiencies of the target (Neo) and reference (Cygb) must be approximately equal. The PCR efficiency of the transgene (Figure 31) and of Cygb (Dr. Aswin Mangerich, Doctoral thesis, University of Konstanz) was determined by a standard curve with 5-fold dilution steps (0.8, 4, 20 and 100 ng template DNA) and estimated by the following equation:

$$E = (10^{-1/slope} - 1) \times 100$$

As seen in Figure 31, the PCR efficiency of neomycin was nearly 100% (97.1%), revealing a doubling of the particular PCR amplicon in each PCR cycle. A similar efficiency was achieved by amplification of Cygb (98%; Dr. Aswin Mangerich, Doctoral thesis, University of Konstanz). Therefore, the assumption that the amplification efficiencies of target and reference gene in the PCR are approximately equal and the $\Delta\Delta C_t$ method was used to analyze the transgene copy number.

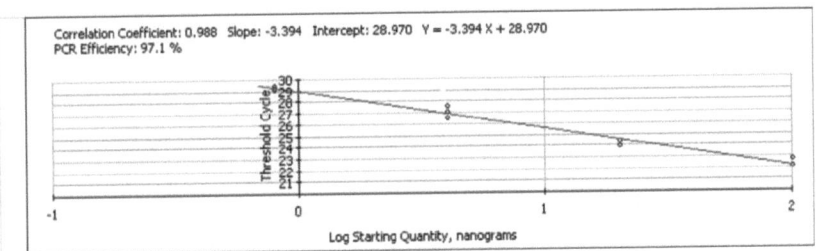

Figure 31: Real-time PCR standard curve for the neomycin cassette of the hPARP-1 transgene. The standard curve was generated by iCycler software from threshold cycles of transgene (Neo) by serial dilution steps of starting material (0.8, 4, 20 and 100 ng template DNA) measured in triplicate.

Results

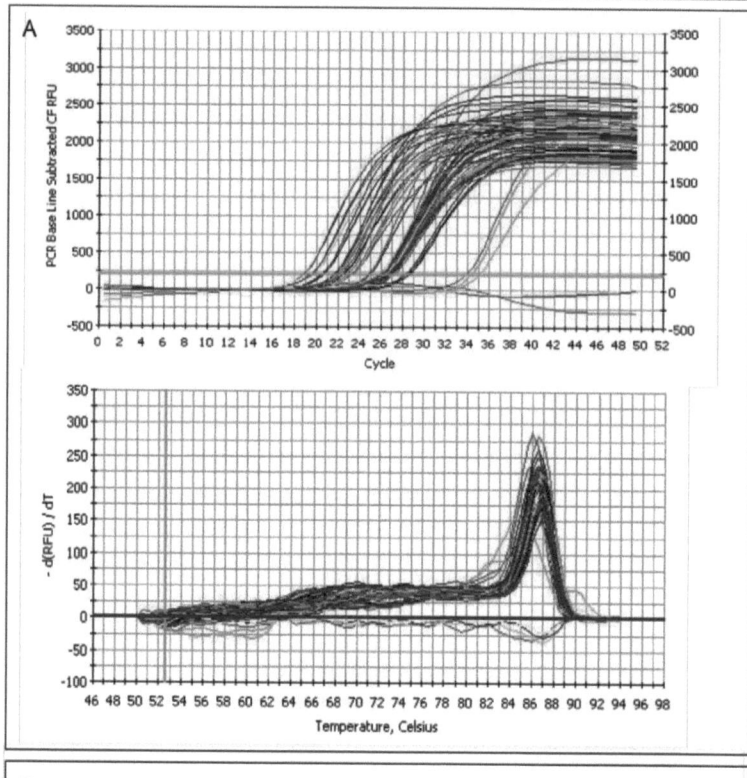

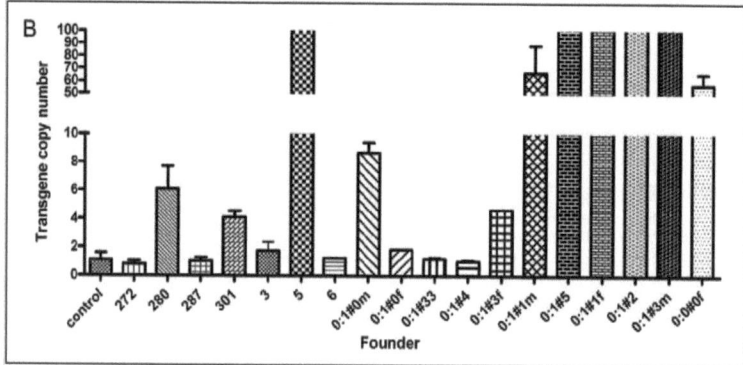

Figure 32: Copy number of *hPARP-1* in transgenic founder mice. (A, top) Representative amplification and (A, bottom) melt-curve profiles with Sybr®Green detection from transgenic founder mice by real-time PCR resulting from amplification with primer sets for *Neo* and *Cygb*. (B) Transgene copy number was determined by relative quantification. From left to right: control, mouse DNA obtained from Dr. Aswin Mangerich with one copy of neomycin gene per genome; 272-301, first microinjection series; 3-6, second microinjection series; 0:1#0m – 0:0#0, third microinjection series. Calculated copy number >100 are not shown.

Results

The quantitative real-time PCR experiments for determination of the transgene copy number in all 18 founder lines were run by using the same primer sets as in the experiment performed to determine PCR efficiency (*Neo*, 50509/50510; *Cygb*, AMa09/AMa10). Each measurement was done in triplicate with 100 ng template DNA (Figure 32; A, top). In order to exclude nonspecific products, a melt-curve analysis was performed immediately after the amplification (Figure 32; A, bottom). The transgene copy number was determined by relative quantitative real-time PCR. Mouse DNA containing only one copy of *Neo* per genome (mouse obtained from Dr. Aswin Mangerich, University of Konstanz) served as control. Copy number exceeding the arithmetical value of 100 are not shown, as small deviations in individual efficiency of amplifications would lead to huge errors in calculation of copy number, especially when the number of threshold cycles in real-time PCR between *Neo* and *Cygb* differs by more than 5. As a result, the melt-curves obtained for each PCR product revealed a single peak at temperatures of approximately 86.5°C for *Neo* and 87°C for *Cygb*. Both peaks represent the specifically amplified products and were not detectable in no-template controls (Figure 32; A, bottom). In the majority of the 18 tested transgenic founder mice, the transgene copy number varied between 1 and 8. Two transgenic mice had a transgene copy number exceeding 50, and five mice had more than 100 copies integrated in their genome (Figure 32; B).

5 Discussion

5.1 Inhibition of PARP-1/-2 in mouse $Parp-1^{+/+}$ and $Parp-1^{-/-}$ fibroblasts

5.1.1 Detection and selective inhibition of PAR formation in $Parp-1^{+/+}$ and $Parp-1^{-/-}$ mouse fibroblasts

For more than 25 years, PARP-1 constitutes a target for the design of appropriate inhibitors, as inhibitors of this enzyme have the potential to enhance the cytotoxicity of DNA-damaging antitumor agents and to exert therapeutic effects in a number of pathological conditions such as diabetes (Pieper *et al.*, 1999a; Akiyama *et al.*, 2001; Soriano *et al.*, 2001; Szabo, 2002; Ferraris *et al.*, 2003; Weseler *et al.*, 2009), inflammation (Szabó and Dawson, 1998), hemorrhagic shock (Liaudet *et al.*, 2000; McDonald *et al.*, 2000; Roesner *et al.*, 2006), myocardial ischemic events (Wayman *et al.*, 2001; Kaplan *et al.*, 2005; Song *et al.*, 2008; Oh *et al.*, 2009), stroke (Abdelkarim *et al.*, 2001; Chiarugi *et al.*, 2003; Ferraris *et al.*, 2003) and other diseases. PARP-1 inhibitors have been successfully evaluated in a number of animal models of reperfusion, degenerative and vascular diseases as well as inflammation (for a review, see: Jagtap and Szabó, 2005)). Recent data also support a contribution of PARP-1 activation to the pathophysiology of diseases with low-grade systemic inflammation, such as chronic obstructive pulmonary disease (COPD) and type-2 diabetes, indicating a potential application of PARP-1 inhibitors for these patient groups (Hageman *et al.*, 2003; Weseler *et al.*, 2009).

Initially, based on the high degree of homology of the PARP catalytic domain between species, it has been suggested that PARP inhibitors might exhibit no difference in terms of potency in human, rat, and mouse tissues (de Murcia *et al.*, 1994; Iwashita *et al.*, 2004b; Iwashita *et al.*, 2004c; Kinoshita *et al.*, 2004), and it was speculated that none of the PARP inhibitors existing at that time would be able to discriminate between PARP-1 and PARP-2 (Oliver *et al.*, 2004). However, Perkins (Perkins *et al.*, 2001) discovered compounds of the quinazolinone and phthalazinone structure with modest selectivity for PARP-1 and PARP-2, respectively. Distinct binding modes necessary for discrimination between ligands and each isoenzyme have then been identified, enabling the synthesis of quinazolinones (e.g., FR247304), with selectivity for PARP-1, and quinoxalines (e.g., FR261529), with selectivity for PARP-2 (Iwashita *et al.*, 2004a; Iwashita *et al.*, 2004b; Ishida *et al.*, 2006), thus demonstrating the feasibility of designing PARP-isoform selective ligands. We recently characterized four imidazoquinolinone, imidazopyridine and isoquinolindione derivatives with PARP inhibitory properties: among these compounds, the isoquinolindione BYK204165 was found to be 100-fold more selective for PARP-1 (Eltze *et al.*, 2008). In terms of selectivity for PARP-1, BYK204165 outperforms that of recently reported quinazolinones, such as FR247304, with 10- to 39-fold selectivity for PARP-1 over PARP-2 (Iwashita *et al.*, 2004b; Ishida

Discussion

et al., 2006), and to our knowledge BYK204165 is the most selective presently available. Also very recently, a series of isoquinolinone derivatives, e.g. UPF-1069, with selective PARP-2 inhibiting properties have been described (Pellicciari et al., 2008; Moroni et al., 2009). The use of inhibitors being highly selective for PARP-1 and PARP-2 might provide an important tool to dissect cellular functions mediated by one isoform alone or both. However, at present it is far from being clear, whether or not selective inhibition of either PARP-1 or PARP-2 might constitute a valuable therapeutic strategy for any disease; nevertheless, such compounds interacting solely with one isoform might give new insights into possibly different roles of the two isoforms in the cell.

In order to confirm the 100-fold selectivity and the efficacy of the new inhibitor BYK204165 for PARP-1 compared to PARP-2 in a cellular system, this compound was evaluated in $Parp\text{-}1^{-/-}$ mouse 3T3 fibroblasts, where the only DNA damage-induced cellular PAR formation after treating the cells with H_2O_2 results from PARP-2 activity. In previous experiments, BYK204165 was tested in a cell-free system with recombinant human PARP-1 (pIC_{50} 7.35) and, because at the time the experiments were started, human PARP-2 enzyme was not available, with recombinant murine PARP-2 (pIC_{50} 5.38) (Eltze et al., 2008). $Parp\text{-}1^{-/-}$ and $Parp\text{-}1^{+/+}$ mouse 3T3 fibroblasts were treated either with BYK204165 or with BYK236864, the latter displaying no selectivity for PARP-1 (pIC_{50} 7.81) or PARP-2 (pIC_{50} 7.43). Both compounds were investigated for their potency to inhibit PAR formation detected by immunofluorescence analysis upon stimulation of the cells with H_2O_2. The results obtained thereof clearly demonstrate that the unselective inhibitor, BYK236864, completely abrogates PAR immunostaining in the cell nuclei at 3 µM and above in $Parp\text{-}1^{+/+}$ mouse 3T3 fibroblasts (Figure 7). In contrast, the PARP-1 selective inhibitor, BYK204165, does not inhibit PAR formation completely, even at 10 µM (Figure 7), confirming the formerly detected PARP-1 selectivity of BYK204165 with cell-free recombinant PARP-1/-2 enzymes in a cellular assay. To prove whether the residual PAR formation in $Parp\text{-}1^{+/+}$ cells treated with BYK204165 at 3 µM and above was solely due to a remaining PARP-2 activity, similar experiments were performed in $Parp\text{-}1^{-/-}$ mouse 3T3 fibroblasts (Figure 9). Due to the fact that PARP-2 accounts for 5 to 10% of the total, maximally stimulated PARP activity (Shieh et al., 1998; Amè et al., 1999; Schreiber et al., 2002), these cells were treated with a 10-fold higher concentration of H_2O_2 (50 mM in $Parp\text{-}1^{-/-}$ vs. 5 mM in $Parp\text{-}1^{+/+}$), however, as expected, PAR formation was still much weaker compared to wt cells (Figure 9, A vs. Figure 7, B). The PARP-1 selectivity of BYK204165 was again confirmed, as PAR formation was not diminished even at 3 µM of the compound, but already completely abrogated at 0.3 µM and above with the unselective BYK236864 in $Parp\text{-}1^{-/-}$ cells (Figure 9, C - D). The still persisting, soft and nonspecific cytoplasmic background staining at 10 µM with BYK236864 (Figure 8, F), is a typical phenomenon and easily distinguishable from the

Discussion

granular pattern of PAR formation in nuclei upon DNA-damaging treatment of the cells with either no or low concentrations of PARP inhibitors (Figure 8, B and C).

Taken together, the novel compound BY204165 might be a valuable tool to selectively inhibit functions of PARP-1 in biological systems, thereby leaving those of PARP-2 untouched. This unique property makes it an ideal drug to dissect the different roles of both PARP isoforms in modulating cellular responses to DNA damage.

5.2 DNA repair and viability in hPARP-1 overexpressing rodent cells

5.2.1 Toxicity induced by alkylating agents in hPARP-1 overexpressing COMF10 cells

In order to study the influence of hPARP-1 overexpression on the cellular responses after DNA damage by alkylating agents, Chinese hamster COMF10 cells with Dex-inducible overexpression of hPARP-1 were treated with either MNNG or MMS. Measurement of necrosis, apoptosis and cell viability was used as experimental endpoint. As a result, Dex-induced hPARP-1 overexpression significantly increased the fraction of necrotic cells and decreased the fraction of viable cells after treatment with MMS at ≥ 500 µM, compared to COMF10 cells without Dex ($p < 0.001$) (Figure 10, A and B). Likewise, this dramatic increase in the necrotic fraction could also be observed after exposing the cells to MNNG between 12.5 and 20 µM ($p < 0.01$ to 0.001), whereas the fraction of apoptotic cells nearly completely disappeared in hPARP-1 overexpressing COMF10 cells compared to COMF10 cells without Dex (Figure 10, C and D). On the other hand, when respective experiments were performed with the control cell line COR4 lacking the hPARP-1 expression plasmid, no significant differences in the fractions of necrotic, apoptotic and viable cells after treatment with MMS or MNNG could be observed, thus ruling out any non-specific effects of Dex with respect to necrosis or apoptosis (Figure 11). However, a slight but not significant beneficial impact on cell survival could be detected in MNNG- and Dex-treated COR4 cells compared to respective Dex-untreated controls, indicating a slight non-specific effect of Dex on this particular endpoint (Figure 11, C and D). The stronger decrease in overall survival rate observed in hPARP-1 overexpressing COMF10 cells compared to Dex-untreated cells, confirms earlier results, demonstrating that MNNG led to a reduction in survival of hPARP-1 overexpressing COMF10 cells, in this case determined by colony-forming assay (Meyer *et al.*, 2000).

Since PARP-1 overactivation can lead to energy starvation by NAD^+, and consequently ATP depletion, one could speculate that the strong decrease in the number of viable cells and the concomitant increase of necrotic cell fraction in hPARP-1 overexpressing cells after damage with MNNG or MMS, could be ascribed to energy starvation. Furthermore, hPARP-1 overexpressing cells are able to produce much more PAR, particularly due to a 5-fold overexpression of hPARP-1 protein and the increased enzyme activity between exogenous human PARP-1 and endogenous hamster PARP-1 (Grube and Bürkle, 1992; Meyer et al., 2000). Previous data have shown that overexpression of hPAPR-1 in COMF10 cells increases PAR-formation after γ-irradiation or MNNG treatment. However, the levels of NAD^+ and ATP in CO60-derived hamster cells constitutively overexpressing PARP-1 did not significantly decrease after γ-irradiation measured at time points of 10, 20, 40 and 120 min (Van Gool et al., 1997; Meyer et al., 2000). However, in mouse embryonic fibroblasts (MEFs), it was demonstrated that MNNG elicited a concentration-dependent decrease in NAD^+ and ATP in wild type cells, but also in MEFs lacking the central apoptotic mediators bax and bak, which may explain necrotic cell death (Zong et al., 2004). Moreover, the same group could show that cells undergo necrosis after treatment with alkylating agents only during aerobic glycolysis to maintain their bioenergetic state, mostly in case of dividing cells, when they are compromised by rapid ATP depletion and forced to cell death in response to PARP activation. Therefore, the most likely explanation is that necrosis of the fast dividing COMF10 cells overexpressing human PARP-1 is due to a bioenergetic catastrophe. Furthermore, it was hypothesized previously that the amount of polymer formed plays a crucial role in determining either sensitivity or resistance to genotoxic stress, and that an optimal level of damage-induced cellular PAR exists to ensure cell survival (Bürkle, 2001). As PARP-1 is implicated in various basic cellular mechanisms, such as cell cycle control (Cohen-Armon, 2007; Carbone et al., 2008) and transcriptional regulation (Wacker et al., 2007), it is necessary that the effective level of poly(ADP-ribosyl)ation is strictly controlled, as it is absolutely critical as a prerequisite for cells to remain functionally intact and viable.

5.2.2 Expression of hPARP-1 in murine lymphoma EL-4 cells

As PARP-1 has been shown to act as a negative regulator of genomic instability, the effects of hPARP-1 overexpression in X-irradiated murine T-cells (EL-4) were analyzed with respect to DNA repair kinetics. For this purpose, EL-4 cells were transiently transfected with an hPARP-1 expression plasmid (pwpt-*hPARP-1*), the optimal transfection conditions being analyzed by hPARP-1 expression and determined by FACS measurements with hPARP-1 specific antibody, which resulted in transfection efficiency of 66% (Figure 12). Interestingly, the transfection efficiency was dependent on the amount of transfection reagent used and was increased by lowering the amount of input DNA (Figure 12, except column 7 and 8).

Discussion

5.2.3 DNA repair kinetics in EL-4 cells treated with jetPEI™

In order to exclude an influence of the transfection reagent jetPEI™ on cellular DNA repair capacity or a toxic effect, jetPEI™ treated EL-4 cells were X-irradiated with 7 Gy, thereafter, DNA repair kinetics were assessed by FADU technique and compared to untreated controls. The transfection reagent neither had any influence on basal DNA strand breaks (Figure 13, P_0), nor an additional effect on DNA damage induction upon X-irradiation (Figure 13, 0 min). Moreover, the DNA repair kinetics remained unaffected by transfection reagent (Figure 13, 10 – 50 min).

5.2.4 Induction of DNA strand breaks by X-irradiation in EL-4 cells

To measure the dose-response relationship between X-irradiation and DNA-damage in EL-4 cells assessed by a DNA-unwinding assay, various X-irradiation doses up to 16 Gy were applied. Thereafter, the fluorescence signal intensity as a measure for the amount of intact double-stranded DNA was detected. A dose-dependent unwinding of DNA in response to X-irradiation could be observed, revealing the dose-dependent formation of DNA strand breaks (Figure 14, A). The higher the X-irradiation dose the higher the amount of DNA strand breaks suffered in the cell, noticeable by a decrease in fluorescence signal intensity. However, the dose-response relationship did not fit a linear curve approximation over the whole range of X-irradiation doses, especially at those >8 Gy. Therefore, for the following cellular DNA repair experiments, three X-irradiation doses within the linear phase (5, 7 and 9 Gy) of the dose-response relationship were used.

5.2.5 Determination of the optimal X-irradiation dose for DNA repair measurements

In order to determine an appropriate X-irradiation dose that induced sufficient DNA damage, detectable by a significant drop in the fluorescence signal compared to un-irradiated control P_0 (endogenous DNA damage/DNA damage induced by experimental procedure), the cells were irradiated at three different doses (5, 7 and 9 Gy) and then further incubated for different times up to 50 min to allow DNA repair (Figure 15). However, it must be kept in mind that application of high X-irradiation doses will inevitably lead to early induction of cell death that would stop the DNA repair process. The three X-irradiation doses used, caused a drop in fluorescence signal of approximately 40% (5 Gy), 50% (7 Gy) and 60% (9 Gy) compared to the un-irradiated control P_0. Moreover, all cells were able to totally recover from suffered DNA damage within 50 min to values even above P_0. The reason that the cells were able to repair DNA to values (120%) exceeding P_0 levels, may partly be due to the application of a rather preliminary experimental protocol prior to its optimization in subsequent studies. In the experiments described here, cells were treated as delineated in "Material and methods" (see section 3.2.1.1), with exception that the 2-h recovery step

Discussion

was missing before X-irradiation. The experimental procedure possibly could not prevent DNA strand breaks during incubation (approximately 30 min), by which time the cells were not kept under optimal assay conditions. Particularly, the cells were processed at 18°C, centrifuged twice and repeatedly pipetted, which eventually induced stress and, as a consequence, led to DNA damage. Actually, after protocol optimization the cells showed both higher P_0 fluorescence signals (data not shown), indicating less endogenous DNA damage, and never generated calculated gray equivalents after X-irradiation and DNA repair surmounting 0 Gy (Figure 16 and Figure 18).

5.2.6 Repair kinetics of DNA strand breaks in EL-4 cells with hPARP-1 expression

In order to explore if hPARP-1 overexpression can alter DNA repair kinetics, El-4 cells were transfected with hPARP-1 expression plasmid pwpt-*hPARP-1*, then exposed to a dose of 7 Gy X-irradiation, and thereafter investigated for their time-dependent DNA repair measured by FADU technique. PARP-1 overexpression had a significant positive impact on DNA repair capacity, as it accelerated DNA repair at each time point (except 15 min) compared to control cells (Figure 16). Additionally, the different shapes of the curves obtained suggest different repair kinetics in the hPARP-1 overexpressing cells, although after 40 min the overall level of repaired DNA in both cell types reaches values comparable to un-irradiated controls (P_0). Particularly, in the control cells, an increase in Gray equivalents occurred between 20 and 30 min, whereas an early slowdown in gray equivalent decrease was observable in the hPARP-1 overexpressing cells between 10 and 15 min. This temporary increase in gray equivalents after 20 and 30 min in control EL-4 cells could be observed in each experiment. This phenomenon was also detectable in human peripheral blood mononuclear cells (PBMCs) after 50 min repair time (unpublished data), which points to the possibility that continuous DNA repair temporarily induced strand breaks. As the repair capacity is slowed in human PBMCs (e.g. after 3.8 Gy X-irradiation, the repair time necessary to reach again P_0 values again amounts to 80 min, Diploma thesis Rebecca Steinhaus, 2008, University of Konstanz), one could speculate that the early increase in gray equivalents in EL-4 cells might have the same reason as the increase observed after 50 min in human PBMCs. X-irradiation can evoke direct breakage of hydrogen or sugar-phosphate bonds in the DNA, and indirectly by formation of free radicals, leading to SSBs and DSBs that are predominantly repaired by the BER pathway. Nevertheless, some of these DNA damages obviously are not recognized and repairable by the early initiation of BER machinery, and thus have to be repaired by a later initiation of NER machinery (Kuraoka *et al.*, 2000). Therefore, one could hypothesize that the specific increase in gray equivalent signal observed between 10 and 15 min marks a switch in repair pathways, *i.e.* from BER towards NER. Particularly, endonucleases within the NER pathway cleave 3′ and 5′ of a DNA damage, leaving a 24-32 bp gap of single-stranded DNA afterwards being re-filled by desoxy-nucleotide insertion using DNA polymerase δ or ε (Hoeijmakers, 2001). To address this hypothesis, an appropriate method would be to inhibit the NER pathway, e.g. by pretreatment with aphidicolin, a DNA-polymerase α and δ inhibitor, which would inhibit the gap filling activity of the polymerase within the NER pathway, thus resulting in accumulation of DNA strand breaks due to enzymatic incisions at DNA damage sites (Collins *et al.*, 1982). Preliminary experiments with human PBMCs, which were UV-C-irradiated and treated with aphidicolin, suffered DNA damage solely repairable by the NER pathway, demonstrating that the amount of double-stranded DNA following damage induction and the ability of the cells to repair DNA, decreased up to 50 min and remained unrepaired up to 120 min. In contrast, cells not treated with aphidicolin, displayed a similar initial decrease but started to increase the amount of double-stranded DNA at 50 min and afterwards (Diploma thesis, Rebecca Steinhaus, 2008, University of Konstanz). These results support the hypothesis of an elevated activity of endonucleases within the first 50 min, and an overcoming polymerase activity afterwards. Taken together, it is suggested that the increase of gray equivalents in both cell lines occurs in a time frame, when most of the DNA damages are repaired by the BER machinery, together with ongoing elevated activity of endonucleases in the NER pathway.

5.2.7 Perturbation of PARP activity by PARP inhibition with PJ34 in EL-4 cells

For the impairment of repair kinetics after DNA damage evoked by X-irradiation in EL-4 cells, the unselective PARP inhibitor PJ34 was used. PAR formation after 40 Gy X-irradiation was detected with a specific antibody (10H) and analyzed by immunofluorescence microscopy. As shown in Figure 17, in the absence of PJ34 most of the cells showed a high PARP activity after X-irradiation, whereas in the presence of 1 μM PJ34, PARP activity was markedly reduced and completely abolished at 5 μM of the inhibitor. In these experiments, the X-irradiation dose applied was high enough to evoke maximal PAR formation, however, about 50% of the cells revealed only weak PAR formation, which was also seen in cells treated with 20 Gy or less (data not shown).

5.2.8 Repair kinetics of DNA strand breaks in EL-4 cells after PARP inhibition

PARP-1 is implicated in genomic stability and participates in different DNA repair mechanisms, mainly in the BER pathway. To determine the effect of PARP-1 inhibition on DNA repair kinetics, EL-4 cells were first treated with the PARP inhibitor PJ34 (5 μM) and then X-irradiated. Thereafter, their ability to repair damaged DNA was measured at different time points between 0 and 40 min and compared to control EL-4 cells without inhibitor (Figure 18). Over the total observation period

Discussion

of 40 min, the DNA repair capacity in PARP inhibited cells was significantly lower than in respective control cells ($p < 0.01$; two-way ANOVA), and also significantly different at distinct time points, e.g. after 5 min and 10 min ($p < 0.01$; Bonferroni posttest). Although PARP activity in the presence of the high concentration of 5 µM PJ34 should be completely inhibited, the cells still exerted a residual but markedly reduced DNA repair capacity compared to control cells without inhibitor. As the progression of both curves is very similar in shape, it appears that both are subjected to similar or identical DNA repair pathway(s), however, it would warrant further studies to explain the common delay in DNA repair, particularly observed in PJ34 treated and untreated cells after 15 min. Interestingly, the extent of initial DNA damage was higher in PJ34 treated cells compared to control cells (Figure 18, 0 min). *Vice versa*, the initial DNA damage was lower in hPARP-1 overexpressing cells compared to control (Figure 16, 0 min). As the cells were kept on ice after X-irradiation for 40 min, presumably these cells were still capable to initiate DNA repair at a low level.

5.3 Experiments for generating *hPARP-1* transgenic mice

5.3.1 Generation of *hPARP-1* transgene for DNA microinjection and its functional expression analysis

To generate transgenic mice with hPARP-1 overexpression in T-cells, several classical cloning steps were necessary for generating an overexpression construct suitable for DNA microinjection into mouse zygotes at the one-cell stadium. The complete overexpression construct (pUCTE5) was characterized to be correct in sequence (sequenced). It was composed of the human ubiquitin C promoter, a floxed neomycin resistance cassette with a transcriptional Stop sequence, the *hPARP-1* cDNA, and finally a 3´ untranslated region of human growth hormone (hGH) minigene carrying a transcription terminator and a polyadenylation signal sequence (Figure 20). Afterwards, functionality of the transgene was assessed in a transfection approach using CO60 hamster cells (Figure 21). As expected, hPARP-1 protein expression could be detected after co-transfection of the cells with pUCTE5 and *Cre* recombinase expression plasmid (turbo-*Cre*), as Cre-mediated excision of the transcriptional *Neo*/Stop sequence allowed expression of the hPARP-1 protein (Figure 21, k). *Vice versa*, expression of hPARP-1 was absent in cells solely being transfected with pUCTE5 but devoid of Cre recombinase (Figure 21, h). However, only approximately 10 - 15% of co-transfected cells were found to express hPARP-1 protein, mirrored by the observation that a high Cre recombinase expression was only detectable in approximately 30% of these cells, thus resulting in a low co-transfection efficiency of 15% (Figure 21, k and l). This extent of expression could be confirmed by Western blot experiments, revealing that hPARP-1 expression in co-transfected CO60

Discussion

cells was only weak compared to the massive constitutive hPARP-1 overexpression in COPF5 cells (Figure 22). However, one should take into consideration that the expression *in vitro* expectedly may be lower, in contrast to the expectedly high expression of hPARP-1 *in vivo*, as in the latter case all T-cells carry the transgene and express Cre recombinase. Taken together, these results show that the transgene is functionally active *in vitro*, verified by immunofluorescence detection and Western blot analysis of the expressed hPARP-1 protein.

5.3.2 Detection of *hPARP-1* transgenic founder mice by real-time PCR

After the evidence for a functional activity of the *hPARP-1* transgene *in vitro*, mouse zygotes were microinjected with the *hPARP-1* transgene DNA. Thereafter, tail biopsies were taken from the newborn mice, which were analyzed for successful incorporation of the transgene into their genome. The first series of microinjections resulted in a low microinjection efficiency, as only 4 of a total number of 79 born mice analyzed by real-time PCR from tail biopsy DNA were characterized as transgenic. It has been shown that under optimal conditions, about 25% of born mice (approximately being equivalent to 20 transgenic mice in this study) that originate from DNA microinjected zygotes, integrate one or more copies of the foreign DNA (Brinster *et al.*, 1985). The reason for the low rate of transgene integration obtained in the first series of DNA microinjection is largely unknown but possibly could rely on, e.g. suboptimal pH and/or other disturbing conditions of the microinjection buffer, and impure or partly degraded DNA. However, the injected DNA material proved to be non-toxic, as in control experiments, zygotes injected with *hPARP-1* transgene and incubated in appropriate medium *in vitro* developed normally to the blastocyst stage. Mice bearing the *hPARP-1* transgene were easily distinguishable from non-transgenic mice in real-time PCR experiments using DNA samples isolated from tail biopsies, whereby the former being typically characterized by a rapid and early increase in fluorescence signal between cycle 22 and 24, (Figure 23, A). The quality of template DNA from all samples proved to be constantly good and was routinely validated in parallel by amplifying the "housekeeping gene" β-actin as a control (Figure 23, B).

5.3.3 Detection of *hPARP-1* transgenic founder mice by conventional PCR

After several matings of the transgenic founder mice that emerged from the first DNA microinjection series, no newborn mice with hPARP-1 expression could be identified. Therefore, a second attempt of DNA microinjection was made, but again with a negative result, because offspring could be obtained only from 3 mice, most of them were cannibalized except for 7 pups. In a third series of DNA microinjection further 61 pups were born, and finally 3 mice from the second and 11 mice from the third series were identified as transgenics. In order to economize the

Discussion

genotyping procedure, the protocol for DNA isolation from tail biopsies was modified, accordingly, the putative transgenic mice were analyzed by conventional PCR technique, which after a slight optimization turned out to work as effective as the procedure using the real-time PCR protocol. Positive transgenic mice could be detected with an hPARP-1 specific primer set that resulted in a 714 bp amplicon size. The DNA samples were routinely validated for quality and quantity in parallel with a β-actin primer set that yielded a 98 bp amplicon size (Figure 24).

5.3.4 Analysis of hPARP-1 protein level in transgenic mice

After generation of hPARP-1 transgenic mice, the next aim was to determine the level of hPARP-1 protein expression in the different transgenic founder lines. For this purpose, the transgenic founder animals were mated with transgenic mice that express Cre recombinase in T-cells (*lck-Cre*), thus being suitable for excision of the transcriptional *Neo*/Stop sequence and allowing hPARP-1 transcription to proceed in their offspring. Protein was isolated from thymus or spleen of these newborn mice and analyzed for hPARP-1 content on a Western blot. Unexpectedly, no hPARP-1 protein was detectable in any of the transgenic founder lines (Figure 25), although antibody specificity for hPARP-1 was sufficiently high, noticeable by detection of recombinant hPARP-1 protein as a control (Figure 25, lane PC). Furthermore, expression of Cre recombinase was detectable in all mice, thereby excluding that the lack of hPARP-1 expression is due to the absence of Cre recombinase expression (Figure 25, anti-cre). The different expression levels of Cre recombinase observed in these mice probably arose from their different age (e.g. the highest expression levels were measured at the age of 5 month; data not shown), as well as from varying expression levels in each individual transgenic *lck-Cre* mouse. As it was also possible, that the expression of hPARP-1 in T-cells was weak and below the Western blot detection limit, further experiments were performed using the MACS® Pan T-cell isolation kit in order to remove all non T-cells from thymus and spleen, because both organs contain other cell types devoid of Cre recombinase expression. Although this procedure raised T-cell purity up to 98% in thymus and up to 93% in spleen, no hPARP-1 expression was detectable in these purified T-cells by Western blot analysis (data not shown). Further experiments were performed with offspring mice from *hPARP-1* transgenic founder animals mated with ubiquitously Cre recombinase expressing mice (*EIIa-Cre*), a measure that should lead to an overall hPARP-1 expression. However, as shown in Figure 26, an expression of hPARP-1 protein in spleen or kidney was lacking. It has been demonstrated earlier, that the level of endogenous PARP-1 expression in mouse kidney is very low (Ogura *et al.*, 1990), therefore, the detection of murine PARP-1 with CII-10 antibody was successful in spleen but

Discussion

remained below the detection limit in kidney. In summary, it was not possible to detect any hPARP-1 protein expression in all *hPARP-1* transgenic mice generated.

5.3.5 Excision of the *Neo*/Stop sequence in genomic DNA of *hPARP-1* transgenic mice

A possible reason for the negative outcome of hPARP-1 expression analysis in mice could be that the excision of the transcriptional *Neo*/Stop sequence did not operate effectively. In order to confirm the successful excision of the *Neo*/Stop sequence *in vivo*, which was shown prior to the experiments *in vitro*, *hPARP-1* transgenic mice were mated with T-cell specific Cre recombinase expressing mice (*lck-Cre* x *hPARP-1*). Thereafter, thymus DNA from their offspring was isolated and used for analysis in a flanking PCR reaction. Methodically, a primer set flanking the *Neo*/Stop region was used (Figure 27, A). As a result, most of the transgenic mice revealed a complete excision of the transgene *in vivo*, demonstrating that Cre recombinase was i) expressed in these mice, and ii) able to excise the transcriptional *Neo*/Stop sequence in order to allow transcription of *hPARP-1* to proceed (Figure 27, B). The primer set used was of high specificity, as thymus DNA from wt mice, which did not contain the transgene, did not generate a detectable amplicon. In these assays, DNA isolated from transgene transfected EL-4 cells and serving as a control (Figure 27, B; lane P), possibly could be contaminated with DNA from the co-transfected EL-4 cells (Figure 27, B; lane P+*Cre*), noticeable by a weak amplicon size of 650 bp that should occur only in the co-transfected EL-4 cells, but not in the single-transfected EL-4 cells lacking the *Cre* recombinase expression plasmid.

For generating transgenic mice that ubiquitously express hPARP-1, transgenic founder mice were mated with a second transgenic mouse line (*EIIa-Cre*) that expresses Cre recombinase under control of a ubiquitously active promoter. DNA samples prepared from tail biopsies of these mice were analyzed by PCR technique. The excision of the floxed *Neo*/Stop cassette was completely carried out in the same *hPARP-1* transgenic founder lines, previously been mated with *lck-Cre* mice, which also had been shown to have a complete excision of the *Neo*/Stop cassette, as verified in DNA prepared from thymus of these mice (compare Figure 27and Figure 28, red boxes). However, it appears that in most of the transgenic pups the excision was either incomplete or even missing. This could be explained by the high sensitivity of the PCR approach used to detect cells that carry the *Neo*/Stop sequence. In comparison with the previous PCR approach using flanking primers, the primer set in this particular PCR experiment bound within the section that was removed by excision. It is feasible that a small number of cells, in which the excision did not occur, were capable to amplify a specific *Neo* amplicon, whereas a great amount of excised *Neo*/Stop regions would favor the short amplicon but inhibit the long amplicon synthesis in the flanking PCR.

Discussion

Although some of the transgenic mice demonstrated a complete excision of the *Neo*/Stop sequence (Figure 28, B; red boxes), they were devoid of hPARP-1 protein expression as analyzed by Western blot (Figure 26, red boxes).

5.3.6 Analysis of founder mice for complete transgene insertion

In order to ensure a complete transgene insertion into the mouse genome, different PCR primer sets with specific binding at the human ubiquitin C promoter-, *Neo*/Stop-, *hPARP-1* cDNA- and the *hGH* minigene-sequence were used to analyze genomic DNA prepared from tail biopsies. As expected, in any of the analyzed founder mice the complete transgene composed of all structure elements listed before could be detected, indicating that DNA microinjection and integration of the heterologous DNA into the mouse genome was successful, and therefore, could not be responsible for the lack of hPARP-1 protein expression (Figure 29).

5.3.7 Analysis of *hPARP-1* mRNA transcription in transgenic animals and their transgene copy number

After the observation that the transgenic mice did not express a detectable amount of hPARP-1 protein, neither in the thymus after mating with *lck-Cre* mice, nor in spleen or kidney after mating with *EIIa-Cre* mice (both matings leading to offspring with correct excision of the transcriptional *Neo*/Stop sequence by functional Cre recombinase *in vivo*), the next aim was to determine whether transcription of *hPARP-1* had occurred at least to the level of mRNA. For this purpose, RNA was isolated from thymus of transgenic mice with putative *hPARP-1* expression, transcribed into cDNA by reverse transcriptase PCR (RT-PCR) and completed by PCR reaction using specific primers for *hPARP-1* sequence amplification. From all transgenic founder lines, only three mice showed expression of an *hPARP-1* mRNA transcript. In these assays, RNA from a transgenic mouse that derived from an alternative strategy (Mangerich *et al.*, 2009) and with proven expression of hPARP-1 protein (obtained from Dr. Aswin Mangerich, University of Konstanz), was used as a positive control (Figure 30, A). Although all three transgenic mice obtained from three different founder lines convincingly displayed an *hPARP-1* mRNA transcript, the respective hPARP-1 protein was not detectable. Therefore, it may be speculated, that the mRNA transcript was instable and enzymatically degraded before ribosomal protein synthesis could be initiated. Theoretically, there are several reasons that would contribute to a degradation of mRNA. Particularly, oligo(dt) primer were used in the RT-PCR that bind to the 3´poly-adenine tail of the processed RNA. Therefore, pre-mRNA processing issues, such as 5´-capping or mRNA splicing are unlikely, as these steps take place prior to the 3´-end polyadenylation (Maniatis and Reed, 2002; Orphanides

Discussion

and Reinberg, 2002; Erkmann and Kutay, 2004). Additionally, several mRNA surveillance complexes in the nucleus would degrade mRNA species that have not yet completed mRNA processing (Mitchell and Tollervey, 2001; Baker and Parker, 2004). It is feasible that the mature *hPARP-1* mRNA had been yet exported from the nucleus to the cytoplasm, but for unknown reasons, either was not translated at the ribosome, or had synthesized a misfolded protein that is targeted for degradation.

After DNA microinjection into the pronucleus of a fertilized egg, the transgene is integrated at random sites into the host genome in a tandem manner, whereby the number of copies of the transgene obtained in each founder animal may additionally be different. In due consideration that gene expression is generally dependent on its transgene copy number, with the exception that a high copy number tandem integration is thought to cause transgene silencing (Dobie *et al.*, 1997; Tang *et al.*, 2007), the transgene copy number of all founder lines was determined. Serial dilutions of genomic DNA isolated from tail biopsies were measured by quantitative real-time PCR assay, using a neomycin-specific primer set combined with a cytoglobin b primer set as a reference. The PCR efficiency of both primer sets was nearly equal (neomycin 97.1%, cytoglobin b 98%) (Figure 31), thus the $\Delta\Delta C_t$ method was applied to quantify the transgene copy number. In the majority of the tested transgenic founder mice, a copy number between 1 and 8 could be determined, however, the remaining animals bore transgenic copy number up to 100 or even more. It has been reported, that an increase in copy number results in a marked decrease in expression of the transgene accompanied by increased chromatin compaction (Garrick *et al.*, 1998). This is consistent with the detected amounts of mRNA in the founder mice, as the transgenic animals no. 2 and 5 (Figure 30), corresponding to the founder lines 0:1#0f and 0:1#3f (Figure 32), had a low copy number of approximately 2 and 5, respectively. However, as an exception, the transgenic animal from founder line 0:0#0 showed the highest mRNA level (Figure 30, no. 4), but also a high copy number of more than 50 copies per genome (Figure 32).

5.3.8 Reasons for the lack of hPARP-1 expression in transgenic mice

As it was clearly demonstrated that none of the transgenic founder lines had expressed hPARP-1 protein in any of the organs examined (Figure 25 and Figure 26), every endeavor should be undertaken to clarify the underlying reasons. On the basis of the results obtained, it can be excluded that the generated transgene was wrong in sequence, because sequencing of the transgene revealed a 100% sequence homology compared to available published sequences from the databases. Furthermore, transcription of *hPARP-1* cDNA could successfully be performed in T-cells, examined by excision of the *Neo*/Stop sequence *in vitro* and *in vivo*, and accompanied by detectable

Discussion

hPARP-1 mRNA transcripts (Figure 27, Figure 28 and Figure 30). Finally, hPARP-1 protein expression was detected by Western blot analyses in co-transfected CO60 cells *in vitro* (Figure 21). Genomic DNA from three transgenic founder lines was isolated, thereafter, the transgene was sequenced in order to exclude any mutations within the transgene of these mice. Although the sequence was confirmed to be correct, for unknown reasons the complexity *in vivo* may have prevented hPARP-1 expression.

After having set up new sequence analysis software (Geneious) in our lab, an intensive examination of the transgene sequence afforded two artificial ATG start codons 5´ proximal to the intended ATG start codon from *hPARP-1* cDNA. It could be confirmed that the two start codons were accurately located within each *loxP* site. The 34 bp long *loxP* site is composed of two 13 bp palindromes flanking an asymmetric 8 bp long core sequence (5´-ATAACTTCGTATA GCATACAT TATACGAAGTTAT-3´, core sequence written in red), which exhibits directionality of the *loxP* site (Hoess and Abremski, 1984). If two of those *loxP* sites within a molecule are located in the same direction, the Cre recombinase would excise the sequence located in between, and after a ligase-mediated ligation, only one *loxP* site would remain. An inversion of the two *loxP* sites orientation would allow the sequence in between to flip around (inverted) (Baubonis and Sauer, 1993). However, the 34 bp *loxP* site can be introduced into the genome in both directions, and the orientation on a plasmid map is depicted as filled triangle, with the tip marking the direction of the *loxP* site (▶, forward; ◀, reverse). As the two ATG start codons were detected in the transgene, it was realized that the reversed *loxP* site orientation probably represents an inevitable pitfall, as it comprises two artificial ATG start codons (5´-ATAACTTCGTATA ATGTATGC TTATACGAAGTTAT-3´, ATG start codons are underlined). This reversed orientation of the *loxP* site was used in plasmid PGKneotpAlox2. The plasmid map published on the website of addgene and received from Philippe Soriano (Seattle, USA), may undoubtedly be incorrect, as it shows the two *loxP* sites depicted in forward orientation. However, as a mistake the orientation is reversed, generating two artificial ATG start codons (Figure 33). Furthermore, the sequence published on the same website contains the reversed *loxP* sites contrary to that in the plasmid map. That is why the sequencing of the transgene partly being composed of PGKneotpAlox2, was thought to be correct, and at first did not arouse suspicion for its failure. Moreover, several publications have shown that the plasmid PKGneotpAlox2 can successfully be used as a floxed transcriptional stop element without major complications (Soriano, 1999; Srinivas *et al.*, 2001; Gargioli and Slack, 2004; Yamamoto *et al.*, 2009). A sequence request from the Yamamoto group and a subsequent sequence analysis revealed that in their group only the PGK promoter and the *Neo* cassette were implemented, whereas the *loxP* sites originated from other plasmids had the correct orientation.

Discussion

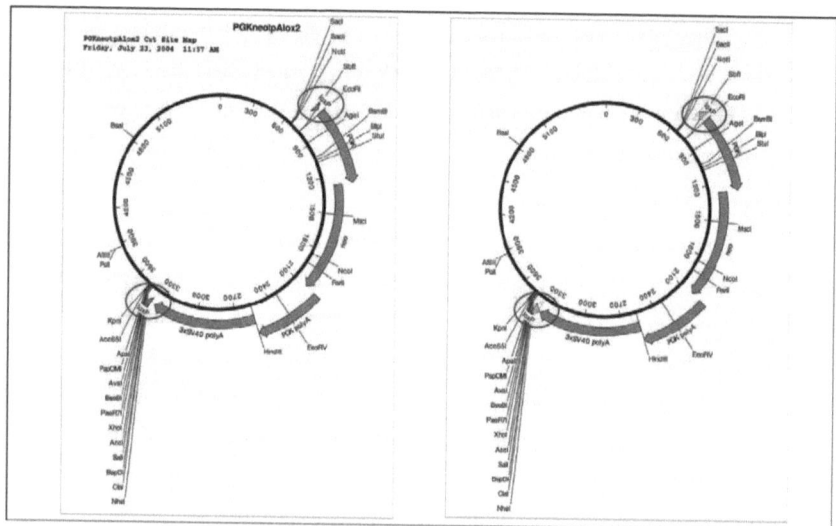

Figure 33: Schematic representation of PGKneotpAlox2. (Left) Plasmid map from addgene (http://www.addgene.org/pgvec1?f=c&cmd=findpl&identifier=13444), legends were adjusted in their position. (Right) Plasmid map with corrected *loxP* orientations. Red circles are focused on the *loxP* sites.

However, not every AUG needs to serve as start codon that is recognized as the transcriptional start site by the ribosome. The strength of the transcriptional start site depends on the Kozak sequence, thereby spanning from 6 nucleotides upstream to 3 nucleotides downstream from adenine of the start AUG referred as number 1 (Harhay et al., 2005). A strong start site is indicated by nucleotide G at position -6, nucleotide A or G at position -3, and G at position +4. The ATG characteristic of the *hPARP-1* cDNA sequence is consistent with the Kozak sequence, whereas the first ATG within the *loxP* site only contains A at position -3, indicating a possible but weaker start site. Additionally, also the second ATG within the *loxP* site is not consistent with the required nucleotides. Only the latter ATG is in frame with the ATG from the ATG start codon of the *hPARP-1* cDNA sequence, and therefore, should lead to its translation into a correct amino acid sequence. Generally, initiation of ribosomal translation starts at the first 5´proximal AUG codon, however, if the first AUG is suboptimally composed and does not mate with the required Kozak sequence, it possibly remains unrecognized, consequently, initiation would start more downstream (Kochetov et al., 2005).

Taken together, these considerations might explain the results obtained, however, in this regard further efforts and data are needed to thoroughly clarify the lack of hPARP-1 protein expression. *In vivo*, the extent of mRNA formation of the *hPARP-1* transgene can be regarded satisfactory,

Discussion

noticeable by the level of *hPARP-1* mRNA detected, nevertheless, a minor fraction that might have undergone an alternative translation start at the AUG of *hPARP-1*, could be below the detection limit. The detectable but low level of hPARP-1 protein *in vitro*, might be due to an increased transgene copy number per cell that would result in a higher level of transgene mRNA associated with a higher and detectable hPARP-1 protein level.

5.4 General summary and discussion

In the present work, a highly selective inhibitor for PARP-1, BYK204165, was characterized for its potency and selectivity to inhibit PARP-1 and PARP-2 mediated PAR formation in 3T3 fibroblasts from *Parp-1$^{+/+}$* and *Parp-1$^{-/-}$* mice in response to oxidative stress. The 100-fold PARP-1 selectivity of the compound was confirmed by its failure to inhibit PARP-2 in both *Parp-1$^{+/+}$* and *Parp-1$^{-/-}$* fibroblasts. Thus, the new compound might provide a novel and convenient functional tool *in vitro* to dissect the contributions of and interactions between PARP-1 and PARP-2 regarding their overlapping and more specific functions. Moreover, the compound may be valuable in *in vivo* experimental set-ups aiming to elucidate the role of both isoenzymes in a number of pathophysiological disease conditions, such as diabetes, inflammation and stroke, in which pharmacological inhibitors of PARP have shown to elicit beneficial effects.

Measurements of apoptosis, necrosis and DNA repair served to explore the consequences of DNA damage induction after treatment with MMS, MNNG or X-irradiation in stably overexpressed human PARP-1 (hPARP-1) Chinese hamster cells (COMF10) and murine lymphoma EL-4 cells, respectively. Analysis of cell viability after treatment with alkylating agents revealed consistently larger fractions of necrotic cells in the COMF10 cells compared to control. DNA repair kinetic measurements after X-irradiation of hPARP-1 overexpressing EL-4 cells demonstrated acceleration in DNA repair, whereas pharmacological inhibition of PARP caused both reduction and delay of cellular DNA repair capacity. The effects observed are consistent with the dual role of PARP in cell death and DNA repair, and underline the decisive momentarily level of PAR either to sensitize cells to cytotoxic effects, or to aid in recovery from DNA damage.

In order to generate an *in vivo* system for tissue-specific overexpression of hPARP-1 protein, transgenic mice were obtained by DNA microinjection of an *hPARP-1* cDNA containing transgene, the transcription of which should be initiated *in vivo* by crossing with transgenic tissue-specific "Cre-deleter" mice. Despite cell culture validation of the expression construct and transgene expression on the level of mRNA, presumably due to an unintentional AUG start codon within a reversed *loxP* site integrated 5′ proximal from the AUG start codon of the *hPARP-1* cDNA, no protein expression of hPARP-1 could be detected in mice. However, as it could be demonstrated that the transgene is able to express hPARP-1 protein *in vitro*, it should need only little efforts to exchange few bases of the core sequence within the *loxP* sequence by site-directed mutagenesis. This technique would remove the unintended AUG start codon and allow to follow-up the designated studies with new DNA microinjections for the generation transgenic mice with tissue-specific overexpression of hPARP-1 protein.

6 References

Abdelkarim, G. E. et al. (2001). "Protective effects of PJ34, a novel, potent inhibitor of poly(ADP-ribose) polymerase (PARP) in in vitro and in vivo models of stroke." Int J Mol Med **7**(3): 255-60.

Adelfalk, C. et al. (2003). "Physical and functional interaction of the Werner syndrome protein with poly-ADP ribosyl transferase." FEBS Lett **554**(1-2): 55-8.

Akiyama, T. et al. (2001). "Activation of Reg gene, a gene for insulin-producing beta -cell regeneration: poly(ADP-ribose) polymerase binds Reg promoter and regulates the transcription by autopoly(ADP-ribosyl)ation." Proc Natl Acad Sci U S A **98**(1): 48-53.

Althaus, F. R. (1992). "Poly ADP-ribosylation: a histone shuttle mechanism in DNA excision repair." J Cell Sci **102** (Pt 4): 663-70.

Altmeyer, M. et al. (2009). "Molecular mechanism of poly(ADP-ribosyl)ation by PARP1 and identification of lysine residues as ADP-ribose acceptor sites." Nucleic Acids Res **37**(11): 3723-38.

Alvarez-Gonzalez, R. et al. (1999). "Selective loss of poly(ADP-ribose) and the 85-kDa fragment of poly(ADP-ribose) polymerase in nucleoli during alkylation-induced apoptosis of HeLa cells." J Biol Chem **274**(45): 32122-6.

Amé, J. C. et al. (2009). "Radiation-induced mitotic catastrophe in PARG-deficient cells." J Cell Sci **122**(Pt 12): 1990-2002.

Amè, J. C. et al. (1999). "PARP-2, a novel mammalian DNA damage-dependent poly(ADP-ribose) polymerase." J Biol Chem **274**(25): 17860-8.

Amè, J. C. et al. (2004). "The PARP superfamily." Bioessays **26**(8): 882-93.

Andrabi, S. A. et al. (2006). "Poly(ADP-ribose) (PAR) polymer is a death signal." Proc Natl Acad Sci U S A **103**(48): 18308-13.

Baker, K. E. and R. Parker (2004). "Nonsense-mediated mRNA decay: terminating erroneous gene expression." Curr Opin Cell Biol **16**(3): 293-9.

Banasik, M. et al. (2004). "The effects of organic solvents on poly(ADP-ribose) polymerase-1 activity: implications for neurotoxicity." Acta Neurobiol Exp (Wars) **64**(4): 467-73.

Bassing, C. H. et al. (2002). "The mechanism and regulation of chromosomal V(D)J recombination." Cell **109 Suppl**: S45-55.

Baubonis, W. and B. Sauer (1993). "Genomic targeting with purified Cre recombinase." Nucleic Acids Res **21**(9): 2025-9.

Bauer, P. I. et al. (1992). "Inhibition of DNA binding by the phosphorylation of poly ADP-ribose polymerase protein catalysed by protein kinase C." Biochem Biophys Res Commun **187**(2): 730-6.

Benchoua, A. et al. (2002). "Active caspase-8 translocates into the nucleus of apoptotic cells to inactivate poly(ADP-ribose) polymerase-2." J Biol Chem **277**(37): 34217-22.

Beneke, R. et al. (2000a). "DNA excision repair and DNA damage-induced apoptosis are linked to Poly(ADP-ribosyl)ation but have different requirements for p53." Mol Cell Biol **20**(18): 6695-703.

Beneke, S. et al. (2000b). "Comparative characterisation of poly(ADP-ribose) polymerase-1 from two mammalian species with different life span." Exp Gerontol **35**(8): 989-1002.

Beneke, S. and A. Bürkle (2004). "Poly(ADP-ribosyl)ation, PARP, and aging." Sci Aging Knowledge Environ **2004**(49): re9.

Beneke, S. and A. Bürkle (2007). "Poly(ADP-ribosyl)ation in mammalian ageing." Nucleic Acids Res **35**(22): 7456-65.

References

Beneke, S. et al. (2008). "Rapid regulation of telomere length is mediated by poly(ADP-ribose) polymerase-1." Nucleic Acids Res 36(19): 6309-17.

Besson, V. C. (2009). "Drug targets for traumatic brain injury from poly(ADP-ribose)polymerase pathway modulation." Br J Pharmacol 157(5): 695-704.

Bowes, J. et al. (1998). "Reduction of myocardial reperfusion injury by an inhibitor of poly (ADP-ribose) synthetase in the pig." Eur J Pharmacol 359(2-3): 143-50.

Bratton, S. B. and G. M. Cohen (2001). "Apoptotic death sensor: an organelle's alter ego?" Trends Pharmacol Sci 22(6): 306-15.

Brinster, R. L. et al. (1985). "Factors affecting the efficiency of introducing foreign DNA into mice by microinjecting eggs." Proc Natl Acad Sci U S A 82(13): 4438-42.

Bryant, H. E. et al. (2009). "PARP is activated at stalled forks to mediate Mre11-dependent replication restart and recombination." EMBO J.

Bürkle, A. (2001). "Poly(APD-ribosyl)ation, a DNA damage-driven protein modification and regulator of genomic instability." Cancer Lett 163(1): 1-5.

Bürkle, A. (2005). "Poly(ADP-ribose). The most elaborate metabolite of NAD+." Febs J 272(18): 4576-89.

Bürkle, A. et al. (2005). "Ageing and PARP." Pharmacol Res 52(1): 93-9.

Caldecott, K. W. (2001). "Mammalian DNA single-strand break repair: an X-ra(y)ted affair." Bioessays 23(5): 447-55.

Caldecott, K. W. (2004). "DNA single-strand breaks and neurodegeneration." DNA Repair (Amst) 3(8-9): 875-82.

Carbone, M. et al. (2008). "Poly(ADP-ribosyl)ation is implicated in the G0-G1 transition of resting cells." Oncogene 27(47): 6083-92.

Chalmers, A. et al. (2004). "PARP-1, PARP-2, and the cellular response to low doses of ionizing radiation." Int J Radiat Oncol Biol Phys 58(2): 410-9.

Chambon, P. et al. (1963). "Nicotinamide mononucleotide activation of new DNA-dependent polyadenylic acid synthesizing nuclear enzyme." Biochem Biophys Res Commun 11: 39-43.

Chen, M. et al. (2004). "Mitochondrial-to-nuclear translocation of apoptosis-inducing factor in cardiac myocytes during oxidant stress: potential role of poly(ADP-ribose) polymerase-1." Cardiovasc Res 63(4): 682-8.

Chiarugi, A. (2002). "Inhibitors of poly(ADP-ribose) polymerase-1 suppress transcriptional activation in lymphocytes and ameliorate autoimmune encephalomyelitis in rats." Br J Pharmacol 137(6): 761-70.

Chiarugi, A. (2005). "Poly(ADP-ribosyl)ation and stroke." Pharmacol Res 52(1): 15-24.

Chiarugi, A. et al. (2003). "Novel isoquinolinone-derived inhibitors of poly(ADP-ribose) polymerase-1: pharmacological characterization and neuroprotective effects in an in vitro model of cerebral ischemia." J Pharmacol Exp Ther 305(3): 943-9.

Chiarugi, A. and M. A. Moskowitz (2003). "Poly(ADP-ribose) polymerase-1 activity promotes NF-kappaB-driven transcription and microglial activation: implication for neurodegenerative disorders." J Neurochem 85(2): 306-17.

Cohausz, O. et al. (2008). "The roles of poly(ADP-ribose)-metabolizing enzymes in alkylation-induced cell death." Cell Mol Life Sci 65(4): 644-55.

Cohen-Armon, M. (2007). "PARP-1 activation in the ERK signaling pathway." Trends Pharmacol Sci 28(11): 556-60.

References

Collins, A. R. et al. (1982). "Inhibitors of repair DNA synthesis." Nucleic Acids Res **10**(4): 1203-13.

Cottet, F. et al. (2000). "New polymorphisms in the human poly(ADP-ribose) polymerase-1 coding sequence: lack of association with longevity or with increased cellular poly(ADP-ribosyl)ation capacity." J Mol Med **78**(8): 431-40.

Csiszar, A. et al. (2005). "Role of oxidative and nitrosative stress, longevity genes and poly(ADP-ribose) polymerase in cardiovascular dysfunction associated with aging." Curr Vasc Pharmacol **3**(3): 285-91.

D'Amours, D. et al. (1999). "Poly(ADP-ribosyl)ation reactions in the regulation of nuclear functions." Biochem J **342** (Pt 2): 249-68.

D'Amours, D. et al. (2001). "Gain-of-function of poly(ADP-ribose) polymerase-1 upon cleavage by apoptotic proteases: implications for apoptosis." J Cell Sci **114**(Pt 20): 3771-8.

Dantzer, F. et al. (2000). "Base excision repair is impaired in mammalian cells lacking Poly(ADP-ribose) polymerase-1." Biochemistry **39**(25): 7559-69.

David, K. K. et al. (2009). "Parthanatos, a messenger of death." Front Biosci **14**: 1116-28.

Davidovic, L. et al. (2001). "Importance of poly(ADP-ribose) glycohydrolase in the control of poly(ADP-ribose) metabolism." Exp Cell Res **268**(1): 7-13.

de la Lastra, C. A. et al. (2007). "Poly(ADP-ribose) polymerase inhibitors: new pharmacological functions and potential clinical implications." Curr Pharm Des **13**(9): 933-62.

de Murcia, G. and J. Menissier de Murcia (1994). "Poly(ADP-ribose) polymerase: a molecular nick-sensor." Trends Biochem Sci **19**(4): 172-6.

de Murcia, G. et al. (1994). "Structure and function of poly(ADP-ribose) polymerase." Mol Cell Biochem **138**(1-2): 15-24.

de Murcia, J. M. et al. (1997). "Requirement of poly(ADP-ribose) polymerase in recovery from DNA damage in mice and in cells." Proc Natl Acad Sci U S A **94**(14): 7303-7.

Desnoyers, S. et al. (1995). "Biochemical properties and function of poly(ADP-ribose) glycohydrolase." Biochimie **77**(6): 433-8.

Dobie, K. et al. (1997). "Variegated gene expression in mice." Trends Genet **13**(4): 127-30.

Docherty, J. C. et al. (1999). "An inhibitor of poly(ADP-ribose) synthetase activity reduces contractile dysfunction and preserves high energy phosphate levels during reperfusion of the ischaemic rat heart." Br J Pharmacol **127**(4): 1518-24.

Duriez, P. J. and G. M. Shah (1997). "Cleavage of poly(ADP-ribose) polymerase: a sensitive parameter to study cell death." Biochem Cell Biol **75**(4): 337-49.

El-Khamisy, S. F. et al. (2003). "A requirement for PARP-1 for the assembly or stability of XRCC1 nuclear foci at sites of oxidative DNA damage." Nucleic Acids Res **31**(19): 5526-33.

Eliasson, M. J. et al. (1997). "Poly(ADP-ribose) polymerase gene disruption renders mice resistant to cerebral ischemia." Nat Med **3**(10): 1089-95.

Eltze, T. et al. (2008). "Imidazoquinolinone, imidazopyridine, and isoquinolindione derivatives as novel and potent inhibitors of the poly(ADP-ribose) polymerase (PARP): a comparison with standard PARP inhibitors." Mol Pharmacol **74**(6): 1587-98.

Endres, M. et al. (1997). "Ischemic brain injury is mediated by the activation of poly(ADP-ribose)polymerase." J Cereb Blood Flow Metab **17**(11): 1143-51.

Erkmann, J. A. and U. Kutay (2004). "Nuclear export of mRNA: from the site of transcription to the cytoplasm." Exp Cell Res **296**(1): 12-20.

References

Fahrer, J. et al. (2007). "Quantitative analysis of the binding affinity of poly(ADP-ribose) to specific binding proteins as a function of chain length." Nucleic Acids Res **35**(21): e143.

Faro, R. et al. (2002). "Myocardial protection by PJ34, a novel potent poly (ADP-ribose) synthetase inhibitor." Ann Thorac Surg **73**(2): 575-81.

Ferraris, D. et al. (2003). "Design and synthesis of poly(ADP-ribose) polymerase-1 (PARP-1) inhibitors. Part 4: biological evaluation of imidazobenzodiazepines as potent PARP-1 inhibitors for treatment of ischemic injuries." Bioorg Med Chem **11**(17): 3695-707.

Filipovic, D. M. et al. (1999). "Inhibition of PARP prevents oxidant-induced necrosis but not apoptosis in LLC-PK1 cells." Am J Physiol **277**(3 Pt 2): F428-36.

Gagne, J. P. et al. (2009). "Proteomic investigation of phosphorylation sites in poly(ADP-ribose) polymerase-1 and poly(ADP-ribose) glycohydrolase." J Proteome Res **8**(2): 1014-29.

Gamble, M. J. and R. P. Fisher (2007). "SET and PARP1 remove DEK from chromatin to permit access by the transcription machinery." Nat Struct Mol Biol **14**(6): 548-55.

Gargioli, C. and J. M. Slack (2004). "Cell lineage tracing during Xenopus tail regeneration." Development **131**(11): 2669-79.

Garrick, D. et al. (1998). "Repeat-induced gene silencing in mammals." Nat Genet **18**(1): 56-9.

Germain, M. et al. (1999). "Cleavage of automodified poly(ADP-ribose) polymerase during apoptosis. Evidence for involvement of caspase-7." J Biol Chem **274**(40): 28379-84.

Ghodgaonkar, M. M. et al. (2008). "Depletion of poly(ADP-ribose) polymerase-1 reduces host cell reactivation of a UV-damaged adenovirus-encoded reporter gene in human dermal fibroblasts." DNA Repair (Amst) **7**(4): 617-32.

Gonzalez-Rey, E. et al. (2007). "Therapeutic effect of a poly(ADP-ribose) polymerase-1 inhibitor on experimental arthritis by downregulating inflammation and Th1 response." PLoS One **2**(10): e1071.

Grube, K. and A. Bürkle (1992). "Poly(ADP-ribose) polymerase activity in mononuclear leukocytes of 13 mammalian species correlates with species-specific life span." Proc Natl Acad Sci USA **89**(24): 11759-63.

Guggenheim, E. R. et al. (2009). "Photoaffinity isolation and identification of proteins in cancer cell extracts that bind to platinum-modified DNA." Chembiochem **10**(1): 141-57.

Hageman, G. J. et al. (2003). "Systemic poly(ADP-ribose) polymerase-1 activation, chronic inflammation, and oxidative stress in COPD patients." Free Radic Biol Med **35**(2): 140-8.

Haider, U. G. et al. (2003). "Resveratrol increases serine15-phosphorylated but transcriptionally impaired p53 and induces a reversible DNA replication block in serum-activated vascular smooth muscle cells." Mol Pharmacol **63**(4): 925-32.

Hakem, R. (2008). "DNA-damage repair; the good, the bad, and the ugly." EMBO J **27**(4): 589-605.

Hakmé, A. et al. (2008). "The expanding field of poly(ADP-ribosyl)ation reactions. 'Protein Modifications: Beyond the Usual Suspects' Review Series." EMBO Rep **9**(11): 1094-100.

Halappanavar, S. S. et al. (1999). "Survival and proliferation of cells expressing caspase-uncleavable Poly(ADP-ribose) polymerase in response to death-inducing DNA damage by an alkylating agent." J Biol Chem **274**(52): 37097-104.

Harhay, G. P. et al. (2005). "Characterization of 954 bovine full-CDS cDNA sequences." BMC Genomics **6**: 166.

References

Harris, J. L. *et al.* (2009). "Aprataxin, Poly-ADP Ribose Polymerase 1 (PARP-1) and Apurinic Endonuclease 1 (APE1) function together to protect the genome against oxidative damage." Hum Mol Genet.

Hart, R. W. and R. B. Setlow (1974). "Correlation between deoxyribonucleic acid excision-repair and life-span in a number of mammalian species." Proc Natl Acad Sci U S A **71**(6): 2169-73.

Hassa, P. O. *et al.* (2005). "Acetylation of poly(ADP-ribose) polymerase-1 by p300/CREB-binding protein regulates coactivation of NF-kappaB-dependent transcription." J Biol Chem **280**(49): 40450-64.

Hassa, P. O. *et al.* (2006). "Nuclear ADP-ribosylation reactions in mammalian cells: where are we today and where are we going?" Microbiol Mol Biol Rev **70**(3): 789-829.

Hassa, P. O. and M. O. Hottiger (2008). "The diverse biological roles of mammalian PARPS, a small but powerful family of poly-ADP-ribose polymerases." Front Biosci **13**: 3046-82.

Hendryk, S. *et al.* (2008). "Influence of 5-aminoisoquinolin-1-one (5-AIQ) on neutrophil chemiluminescence in rats with transient and prolonged focal cerebral ischemia and after reperfusion." J Physiol Pharmacol **59**(4): 811-22.

Herceg, Z. and Z. Q. Wang (1999). "Failure of poly(ADP-ribose) polymerase cleavage by caspases leads to induction of necrosis and enhanced apoptosis." Mol Cell Biol **19**(7): 5124-33.

Heyer, B. S. *et al.* (2000). "Hypersensitivity to DNA damage leads to increased apoptosis during early mouse development." Genes Dev **14**(16): 2072-84.

Hoeijmakers, J. H. (2001). "Genome maintenance mechanisms for preventing cancer." Nature **411**(6835): 366-74.

Hoess, R. H. and K. Abremski (1984). "Interaction of the bacteriophage P1 recombinase Cre with the recombining site loxP." Proc Natl Acad Sci U S A **81**(4): 1026-9.

Homburg, S. *et al.* (2000). "A fast signal-induced activation of Poly(ADP-ribose) polymerase: a novel downstream target of phospholipase c." J Cell Biol **150**(2): 293-307.

Honjo, T. *et al.* (2002). "Molecular mechanism of class switch recombination: linkage with somatic hypermutation." Annu Rev Immunol **20**: 165-96.

Horton, J. K. *et al.* (2008). "XRCC1 and DNA polymerase beta in cellular protection against cytotoxic DNA single-strand breaks." Cell Res **18**(1): 48-63.

Huber, A. *et al.* (2004). "PARP-1, PARP-2 and ATM in the DNA damage response: functional synergy in mouse development." DNA Repair (Amst) **3**(8-9): 1103-8.

Ikejima, M. *et al.* (1990). "The zinc fingers of human poly(ADP-ribose) polymerase are differentially required for the recognition of DNA breaks and nicks and the consequent enzyme activation. Other structures recognize intact DNA." J Biol Chem **265**(35): 21907-13.

Ishida, J. *et al.* (2006). "Discovery of potent and selective PARP-1 and PARP-2 inhibitors: SBDD analysis via a combination of X-ray structural study and homology modeling." Bioorg Med Chem **14**(5): 1378-90.

Iwashita, A. *et al.* (2004a). "A new poly(ADP-ribose) polymerase inhibitor, FR261529 [2-(4-chlorophenyl)-5-quinoxalinecarboxamide], ameliorates methamphetamine-induced dopaminergic neurotoxicity in mice." J Pharmacol Exp Ther **310**(3): 1114-24.

Iwashita, A. *et al.* (2004b). "A novel and potent poly(ADP-ribose) polymerase-1 inhibitor, FR247304 (5-chloro-2-[3-(4-phenyl-3,6-dihydro-1(2H)-pyridinyl)propyl]-4(3H)-quinazo linone), attenuates neuronal damage in in vitro and in vivo models of cerebral ischemia." J Pharmacol Exp Ther **310**(2): 425-36.

References

Iwashita, A. et al. (2004c). "Neuroprotective effects of a novel poly(ADP-ribose) polymerase-1 inhibitor, 2-[3-[4-(4-chlorophenyl)-1-piperazinyl] propyl]-4(3H)-quinazolinone (FR255595), in an in vitro model of cell death and in mouse 1-methyl-4-phenyl-1,2,3,6-tetrahydropyridine model of Parkinson's disease." J Pharmacol Exp Ther **309**(3): 1067-78.

Jagtap, P. and C. Szabó (2005). "Poly(ADP-ribose) polymerase and the therapeutic effects of its inhibitors." Nat Rev Drug Discov **4**(5): 421-40.

Jog, N. R. et al. (2009). "Poly(ADP-ribose) polymerase-1 regulates the progression of autoimmune nephritis in males by inducing necrotic cell death and modulating inflammation." J Immunol **182**(11): 7297-306.

Kaplan, J. et al. (2005). "Inhibitors of poly (ADP-ribose) polymerase ameliorate myocardial reperfusion injury by modulation of activator protein-1 and neutrophil infiltration." Shock **23**(3): 233-8.

Kappes, F. et al. (2008). "DEK is a poly(ADP-ribose) acceptor in apoptosis and mediates resistance to genotoxic stress." Mol Cell Biol **28**(10): 3245-57.

Kauppinen, T. M. (2007). "Multiple roles for poly(ADP-ribose)polymerase-1 in neurological disease." Neurochem Int **50**(7-8): 954-8.

Kauppinen, T. M. et al. (2006). "Direct phosphorylation and regulation of poly(ADP-ribose) polymerase-1 by extracellular signal-regulated kinases 1/2." Proc Natl Acad Sci U S A **103**(18): 7136-41.

Kauppinen, T. M. et al. (2009). "Inhibition of poly(ADP-ribose) polymerase suppresses inflammation and promotes recovery after ischemic injury." J Cereb Blood Flow Metab **29**(4): 820-9.

Kawaichi, M. et al. (1981). "Multiple autopoly(ADP-ribosyl)ation of rat liver poly(ADP-ribose) synthetase. Mode of modification and properties of automodified synthetase." J Biol Chem **256**(18): 9483-9.

Kim, J. H. et al. (2008). "Inflammatory and transcriptional roles of poly (ADP-ribose) polymerase in ventilator-induced lung injury." Crit Care **12**(4): R108.

Kim, J. W. et al. (2000a). "Inhibition of homodimerization of poly(ADP-ribose) polymerase by its C-terminal cleavage products produced during apoptosis." J Biol Chem **275**(11): 8121-5.

Kim, J. W. et al. (2000b). "DNA-binding activity of the N-terminal cleavage product of poly(ADP-ribose) polymerase is required for UV mediated apoptosis." J Cell Sci **113** (Pt 6): 955-61.

Kim, M. Y. et al. (2004). "NAD+-dependent modulation of chromatin structure and transcription by nucleosome binding properties of PARP-1." Cell **119**(6): 803-14.

Kim, M. Y. et al. (2005). "Poly(ADP-ribosyl)ation by PARP-1: 'PAR-laying' NAD+ into a nuclear signal." Genes Dev **19**(17): 1951-67.

Kinoshita, T. et al. (2004). "Inhibitor-induced structural change of the active site of human poly(ADP-ribose) polymerase." FEBS Lett **556**(1-3): 43-6.

Kleine, H. et al. (2008). "Substrate-assisted catalysis by PARP10 limits its activity to mono-ADP-ribosylation." Mol Cell **32**(1): 57-69.

Kochetov, A. V. et al. (2005). "The role of alternative translation start sites in the generation of human protein diversity." Mol Genet Genomics **273**(6): 491-6.

Kofler, J. et al. (2006). "Differential effect of PARP-2 deletion on brain injury after focal and global cerebral ischemia." J Cereb Blood Flow Metab **26**(1): 135-41.

References

Koh, D. W. *et al.* (2004). "Failure to degrade poly(ADP-ribose) causes increased sensitivity to cytotoxicity and early embryonic lethality." Proc Natl Acad Sci U S A **101**(51): 17699-704.

Kolthur-Seetharam, U. *et al.* (2006). "Control of AIF-mediated cell death by the functional interplay of SIRT1 and PARP-1 in response to DNA damage." Cell Cycle **5**(8): 873-7.

Korkmaz, A. *et al.* (2007). "Pathophysiological aspects of cyclophosphamide and ifosfamide induced hemorrhagic cystitis; implication of reactive oxygen and nitrogen species as well as PARP activation." Cell Biol Toxicol **23**(5): 303-12.

Kraus, W. L. (2008). "Transcriptional control by PARP-1: chromatin modulation, enhancer-binding, coregulation, and insulation." Curr Opin Cell Biol **20**(3): 294-302.

Krishnakumar, R. *et al.* (2008). "Reciprocal binding of PARP-1 and histone H1 at promoters specifies transcriptional outcomes." Science **319**(5864): 819-21.

Kupper, J. H. *et al.* (1990). "Inhibition of poly(ADP-ribosyl)ation by overexpressing the poly(ADP-ribose) polymerase DNA-binding domain in mammalian cells." J Biol Chem **265**(31): 18721-4.

Kupper, J. H. *et al.* (1996). "Trans-dominant inhibition of poly(ADP-ribosyl)ation potentiates carcinogen induced gene amplification in SV40-transformed Chinese hamster cells." Cancer Res **56**(12): 2715-7.

Kupper, J. H. *et al.* (1995). "trans-dominant inhibition of poly(ADP-ribosyl)ation sensitizes cells against gamma-irradiation and N-methyl-N'-nitro-N-nitrosoguanidine but does not limit DNA replication of a polyomavirus replicon." Mol Cell Biol **15**(6): 3154-63.

Kuraoka, I. *et al.* (2000). "Removal of oxygen free-radical-induced 5',8-purine cyclodeoxynucleosides from DNA by the nucleotide excision-repair pathway in human cells." Proc Natl Acad Sci U S A **97**(8): 3832-7.

Laemmli, U. K. (1970). "Cleavage of structural proteins during the assembly of the head of bacteriophage T4." Nature **227**(5259): 680-5.

Langelier, M. F. *et al.* (2008). "A third zinc-binding domain of human poly(ADP-ribose) polymerase-1 coordinates DNA-dependent enzyme activation." J Biol Chem **283**(7): 4105-14.

Lazebnik, Y. A. *et al.* (1994). "Cleavage of poly(ADP-ribose) polymerase by a proteinase with properties like ICE." Nature **371**(6495): 346-7.

Le Page, F. *et al.* (2003). "Poly(ADP-ribose) polymerase-1 (PARP-1) is required in murine cell lines for base excision repair of oxidative DNA damage in the absence of DNA polymerase beta." J Biol Chem **278**(20): 18471-7.

Lebel, M. *et al.* (2003). "Genetic cooperation between the Werner syndrome protein and poly(ADP-ribose) polymerase-1 in preventing chromatid breaks, complex chromosomal rearrangements, and cancer in mice." Am J Pathol **162**(5): 1559-69.

Li, B. *et al.* (2004). "Identification and biochemical characterization of a Werner's syndrome protein complex with Ku70/80 and poly(ADP-ribose) polymerase-1." J Biol Chem **279**(14): 13659-67.

Liaudet, L. *et al.* (2000). "Protection against hemorrhagic shock in mice genetically deficient in poly(ADP-ribose)polymerase." Proc Natl Acad Sci U S A **97**(18): 10203-8.

Liaudet, L. *et al.* (2001). "Suppression of poly (ADP-ribose) polymerase activation by 3-aminobenzamide in a rat model of myocardial infarction: long-term morphological and functional consequences." Br J Pharmacol **133**(8): 1424-30.

Lipton, S. A. and P. A. Rosenberg (1994). "Excitatory amino acids as a final common pathway for neurologic disorders." N Engl J Med **330**(9): 613-22.

References

Liu, X. et al. (2008). "Poly (ADP-ribose) polymerase activity regulates apoptosis in HeLa cells after alkylating DNA damage." Cancer Biol Ther **7**(6): 934-41.

Livak, K. J. and T. D. Schmittgen (2001). "Analysis of relative gene expression data using real-time quantitative PCR and the 2(-Delta Delta C(T)) Method." Methods **25**(4): 402-8.

Lonskaya, I. et al. (2005). "Regulation of poly(ADP-ribose) polymerase-1 by DNA structure-specific binding." J Biol Chem **280**(17): 17076-83.

Lord, C. J. and A. Ashworth (2008). "Targeted therapy for cancer using PARP inhibitors." Curr Opin Pharmacol **8**(4): 363-9.

Los, M. et al. (2002). "Activation and caspase-mediated inhibition of PARP: a molecular switch between fibroblast necrosis and apoptosis in death receptor signaling." Mol Biol Cell **13**(3): 978-88.

Mandir, A. S. et al. (2000). "NMDA but not non-NMDA excitotoxicity is mediated by Poly(ADP-ribose) polymerase." J Neurosci **20**(21): 8005-11.

Mangerich, A. et al. (2009). "A caveat in mouse genetic engineering: ectopic gene targeting in ES cells by bidirectional extension of the homology arms of a gene replacement vector carrying human PARP-1." Transgenic Res **18**(2): 261-79.

Maniatis, T. and R. Reed (2002). "An extensive network of coupling among gene expression machines." Nature **416**(6880): 499-506.

Maruyama, T. et al. (2007). "Txk, a member of the non-receptor tyrosine kinase of the Tec family, forms a complex with poly(ADP-ribose) polymerase 1 and elongation factor 1alpha and regulates interferon-gamma gene transcription in Th1 cells." Clin Exp Immunol **147**(1): 164-75.

Masutani, M. et al. (2000). "The response of Parp knock-out mice against DNA damaging agents." Mutat Res **462**(2-3): 159-66.

Masutani, M. et al. (1999). "Function of poly(ADP-ribose) polymerase in response to DNA damage: gene-disruption study in mice." Mol Cell Biochem **193**(1-2): 149-52.

McDonald, M. C. et al. (2000). "Effects of 5-aminoisoquinolinone, a water-soluble, potent inhibitor of the activity of poly (ADP-ribose) polymerase on the organ injury and dysfunction caused by haemorrhagic shock." Br J Pharmacol **130**(4): 843-50.

Menissier de Murcia, J. et al. (2003). "Functional interaction between PARP-1 and PARP-2 in chromosome stability and embryonic development in mouse." EMBO J **22**(9): 2255-63.

Messner, S. et al. (2009). "Sumoylation of poly(ADP-ribose) polymerase 1 inhibits its acetylation and restrains transcriptional coactivator function." Faseb J.

Meyer-Ficca, M. L. et al. (2004). "Human poly(ADP-ribose) glycohydrolase is expressed in alternative splice variants yielding isoforms that localize to different cell compartments." Exp Cell Res **297**(2): 521-32.

Meyer, R. et al. (2000). "Negative regulation of alkylation-induced sister-chromatid exchange by poly(ADP-ribose) polymerase-1 activity." Int J Cancer **88**(3): 351-5.

Meyer, R. G. et al. (2007). "Two small enzyme isoforms mediate mammalian mitochondrial poly(ADP-ribose) glycohydrolase (PARG) activity." Exp Cell Res **313**(13): 2920-36.

Mitchell, P. and D. Tollervey (2001). "mRNA turnover." Curr Opin Cell Biol **13**(3): 320-5.

Miwa, M. et al. (1974). "Purification and properties of glycohydrolase from calf thymus splitting ribose-ribose linkages of poly(adenosine diphosphate ribose)." J Biol Chem **249**(11): 3475-82.

Molinete, M. et al. (1993). "Overproduction of the poly(ADP-ribose) polymerase DNA-binding domain blocks alkylation-induced DNA repair synthesis in mammalian cells." EMBO J **12**(5): 2109-17.

References

Moreno-Villanueva, M. et al. (2009). "A modified and automated version of the 'Fluorimetric Detection of Alkaline DNA Unwinding' method to quantify formation and repair of DNA strand breaks." BMC Biotechnol **9**: 39.

Moroni, F. et al. (2009). "Selective PARP-2 inhibitors increase apoptosis in hippocampal slices but protect cortical cells in models of post-ischaemic brain damage." Br J Pharmacol **157**(5): 854-62.

Moroni, F. et al. (2001). "Poly(ADP-ribose) polymerase inhibitors attenuate necrotic but not apoptotic neuronal death in experimental models of cerebral ischemia." Cell Death Differ **8**(9): 921-32.

Mortusewicz, O. et al. (2007). "Feedback-regulated poly(ADP-ribosyl)ation by PARP-1 is required for rapid response to DNA damage in living cells." Nucleic Acids Res **35**(22): 7665-75.

Muiras, M. L. et al. (1998). "Increased poly(ADP-ribose) polymerase activity in lymphoblastoid cell lines from centenarians." J Mol Med **76**(5): 346-54.

Naura, A. S. et al. (2008). "Post-allergen challenge inhibition of poly(ADP-ribose) polymerase harbors therapeutic potential for treatment of allergic airway inflammation." Clin Exp Allergy **38**(5): 839-46.

Nguewa, P. A. et al. (2005). "Poly(ADP-ribose) polymerases: homology, structural domains and functions. Novel therapeutical applications." Prog Biophys Mol Biol **88**(1): 143-72.

Nosseri, C. et al. (1994). "Possible involvement of poly(ADP-ribosyl) polymerase in triggering stress-induced apoptosis." Exp Cell Res **212**(2): 367-73.

Nusinow, D. A. et al. (2007). "Poly(ADP-ribose) polymerase 1 is inhibited by a histone H2A variant, MacroH2A, and contributes to silencing of the inactive X chromosome." J Biol Chem **282**(17): 12851-9.

Ogura, T. et al. (1990). "Striking similarity of the distribution patterns of the poly(ADP-ribose) polymerase and DNA polymerase beta among various mouse organs." Biochem Biophys Res Commun **172**(2): 377-84.

Oh, K. S. et al. (2009). "A novel and orally active poly(ADP-ribose) polymerase inhibitor, KR-33889 [2-[methoxycarbonyl(4-methoxyphenyl) methylsulfanyl]-1H-benzimidazole-4-carboxylic acid amide], attenuates injury in in vitro model of cell death and in vivo model of cardiac ischemia." J Pharmacol Exp Ther **328**(1): 10-8.

Oka, J. et al. (1984). "ADP-ribosyl protein lyase. Purification, properties, and identification of the product." J Biol Chem **259**(2): 986-95.

Oka, S. et al. (2006). "Identification and characterization of a mammalian 39-kDa poly(ADP-ribose) glycohydrolase." J Biol Chem **281**(2): 705-13.

Okano, S. et al. (2003). "Spatial and temporal cellular responses to single-strand breaks in human cells." Mol Cell Biol **23**(11): 3974-81.

Olabisi, O. A. et al. (2008). "Regulation of transcription factor NFAT by ADP-ribosylation." Mol Cell Biol **28**(9): 2860-71.

Oliver, A. W. et al. (2004). "Crystal structure of the catalytic fragment of murine poly(ADP-ribose) polymerase-2." Nucleic Acids Res **32**(2): 456-64.

Opresko, P. L. et al. (2003). "Werner syndrome and the function of the Werner protein; what they can teach us about the molecular aging process." Carcinogenesis **24**(5): 791-802.

Orphanides, G. and D. Reinberg (2002). "A unified theory of gene expression." Cell **108**(4): 439-51.

References

Ouararhni, K. et al. (2006). "The histone variant mH2A1.1 interferes with transcription by down-regulating PARP-1 enzymatic activity." Genes Dev **20**(23): 3324-36.

Pandita, T. K. and C. Richardson (2009). "Chromatin remodeling finds its place in the DNA double-strand break response." Nucleic Acids Res **37**(5): 1363-77.

Pardo, B. et al. (2009). "DNA repair in mammalian cells: DNA double-strand break repair: how to fix a broken relationship." Cell Mol Life Sci **66**(6): 1039-56.

Pellicciari, R. et al. (2008). "On the way to selective PARP-2 inhibitors. Design, synthesis, and preliminary evaluation of a series of isoquinolinone derivatives." ChemMedChem **3**(6): 914-23.

Perkins, E. et al. (2001). "Novel inhibitors of poly(ADP-ribose) polymerase/PARP1 and PARP2 identified using a cell-based screen in yeast." Cancer Res **61**(10): 4175-83.

Pfeiffer, P. et al. (2004). "Pathways of DNA double-strand break repair and their impact on the prevention and formation of chromosomal aberrations." Cytogenet Genome Res **104**(1-4): 7-13.

Pieper, A. A. et al. (1999a). "Poly(ADP-ribose) polymerase-deficient mice are protected from streptozotocin-induced diabetes." Proc Natl Acad Sci U S A **96**(6): 3059-64.

Pieper, A. A. et al. (1999b). "Poly (ADP-ribose) polymerase, nitric oxide and cell death." Trends Pharmacol Sci **20**(4): 171-81.

Pieper, A. A. et al. (2000). "Myocardial postischemic injury is reduced by poly(ADP-ribose) polymerase-1 gene disruption." Mol Med **6**(4): 271-82.

Pillai, J. B. et al. (2005). "Increased expression of poly(ADP-ribose) polymerase-1 contributes to caspase-independent myocyte cell death during heart failure." Am J Physiol Heart Circ Physiol **288**(2): H486-96.

Piskunova, T. S. et al. (2008). "Deficiency in Poly(ADP-ribose) Polymerase-1 (PARP-1) Accelerates Aging and Spontaneous Carcinogenesis in Mice." Curr Gerontol Geriatr Res: 754190.

Pleschke, J. M. et al. (2000). "Poly(ADP-ribose) binds to specific domains in DNA damage checkpoint proteins." J Biol Chem **275**(52): 40974-80.

Pogrebniak, A. et al. (2003). "Poly ADP-ribose polymerase (PARP) inhibitors transiently protect leukemia cells from alkylating agent induced cell death by three different effects." Eur J Med Res **8**(10): 438-50.

Prasad, R. et al. (2001). "DNA polymerase beta -mediated long patch base excision repair. Poly(ADP-ribose)polymerase-1 stimulates strand displacement DNA synthesis." J Biol Chem **276**(35): 32411-4.

Preiss, J. and P. Handler (1958). "Biosynthesis of diphosphopyridine nucleotide. I. Identification of intermediates." J Biol Chem **233**(2): 488-92.

Rajamohan, S. B. et al. (2009). "SIRT1 promotes cell survival under stress by deacetylation-dependent deactivation of poly (ADP-ribose) polymerase 1." Mol Cell Biol.

Rich, T. et al. (2000). "Defying death after DNA damage." Nature **407**(6805): 777-83.

Roesner, J. P. et al. (2006). "Protective effects of PARP inhibition on liver microcirculation and function after haemorrhagic shock and resuscitation in male rats." Intensive Care Med **32**(10): 1649-57.

Rongvaux, A. et al. (2003). "Reconstructing eukaryotic NAD metabolism." Bioessays **25**(7): 683-90.

Ruscetti, T. et al. (1998). "Stimulation of the DNA-dependent protein kinase by poly(ADP-ribose) polymerase." J Biol Chem **273**(23): 14461-7.

References

Saberi, A. et al. (2007). "RAD18 and poly(ADP-ribose) polymerase independently suppress the access of nonhomologous end joining to double-strand breaks and facilitate homologous recombination-mediated repair." Mol Cell Biol **27**(7): 2562-71.

Saenz, L. et al. (2008). "Transcriptional regulation by poly(ADP-ribose) polymerase-1 during T cell activation." BMC Genomics **9**: 171.

Sastry, S. S. and E. Kun (1990). "The interaction of adenosine diphosphoribosyl transferase (ADPRT) with a cruciform DNA." Biochem Biophys Res Commun **167**(2): 842-7.

Schorpp, M. et al. (1996). "The human ubiquitin C promoter directs high ubiquitous expression of transgenes in mice." Nucleic Acids Res **24**(9): 1787-8.

Schreiber, V. et al. (2002). "Poly(ADP-ribose) polymerase-2 (PARP-2) is required for efficient base excision DNA repair in association with PARP-1 and XRCC1." J Biol Chem **277**(25): 23028-36.

Schreiber, V. et al. (2006). "Poly(ADP-ribose): novel functions for an old molecule." Nat Rev Mol Cell Biol **7**(2): 517-528.

Schreiber, V. et al. (1995). "A dominant-negative mutant of human poly(ADP-ribose) polymerase affects cell recovery, apoptosis, and sister chromatid exchange following DNA damage." Proc Natl Acad Sci U S A **92**(11): 4753-7.

Shall, S. and G. de Murcia (2000). "Poly(ADP-ribose) polymerase-1: what have we learned from the deficient mouse model?" Mutat Res **460**(1): 1-15.

Shieh, W. M. et al. (1998). "Poly(ADP-ribose) polymerase null mouse cells synthesize ADP-ribose polymers." J Biol Chem **273**(46): 30069-72.

Sodhi, R. K. et al. (2009). "Protective effects of caspase-9 and poly(ADP-ribose) polymerase inhibitors on ischemia-reperfusion-induced myocardial injury." Arch Pharm Res **32**(7): 1037-43.

Song, Z. F. et al. (2008). "Inhibition of the activity of poly (ADP-ribose) polymerase reduces heart ischaemia/reperfusion injury via suppressing JNK-mediated AIF translocation." J Cell Mol Med **12**(4): 1220-8.

Soriano, F. G. et al. (2001). "Diabetic endothelial dysfunction: role of reactive oxygen and nitrogen species production and poly(ADP-ribose) polymerase activation." J Mol Med **79**(8): 437-48.

Soriano, P. (1999). "Generalized lacZ expression with the ROSA26 Cre reporter strain." Nat Genet **21**(1): 70-1.

Southan, G. J. and C. Szabó (2003). "Poly(ADP-ribose) polymerase inhibitors." Curr Med Chem **10**(4): 321-40.

Srinivas, S. et al. (2001). "Cre reporter strains produced by targeted insertion of EYFP and ECFP into the ROSA26 locus." BMC Dev Biol **1**: 4.

Stivala, L. A. et al. (2001). "Specific structural determinants are responsible for the antioxidant activity and the cell cycle effects of resveratrol." J Biol Chem **276**(25): 22586-94.

Strosznajder, J. B. et al. (2000). "Age-related alteration of poly(ADP-ribose) polymerase activity in different parts of the brain." Acta Biochim Pol **47**(2): 331-7.

Szabo, C. (2002). "PARP as a Drug Target for the Therapy of Diabetic Cardiovascular Dysfunction." Drug News Perspect **15**(4): 197-205.

Szabó, C. and V. L. Dawson (1998). "Role of poly(ADP-ribose) synthetase in inflammation and ischaemia-reperfusion." Trends Pharmacol Sci **19**(7): 287-98.

Tanaka, Y. et al. (1987). "Poly (ADP-ribose) synthetase is phosphorylated by protein kinase C in vitro." Biochem Biophys Res Commun **148**(2): 709-17.

References

Tang, W. *et al.* (2007). "Genetic transformation and gene silencing mediated by multiple copies of a transgene in eastern white pine." J Exp Bot **58**(3): 545-54.

Tao, Z. *et al.* (2009). "Identification of the ADP-Ribosylation Sites in the PARP-1 Automodification Domain: Analysis and Implications." J Am Chem Soc.

Tewari, M. *et al.* (1995). "Yama/CPP32 beta, a mammalian homolog of CED-3, is a CrmA-inhibitable protease that cleaves the death substrate poly(ADP-ribose) polymerase." Cell **81**(5): 801-9.

Thiemermann, C. *et al.* (1997). "Inhibition of the activity of poly(ADP-ribose) synthetase reduces ischemia-reperfusion injury in the heart and skeletal muscle." Proc Natl Acad Sci U S A **94**(2): 679-83.

Tremblay, M. *et al.* (2009). "Nucleotide excision repair and photolyase repair of UV photoproducts in nucleosomes: assessing the existence of nucleosome and non-nucleosome rDNA chromatin in vivo." Biochem Cell Biol **87**(1): 337-46.

Trucco, C. *et al.* (1998). "DNA repair defect in poly(ADP-ribose) polymerase-deficient cell lines." Nucleic Acids Res **26**(11): 2644-9.

Valdor, R. *et al.* (2008). "Regulation of NFAT by poly(ADP-ribose) polymerase activity in T cells." Mol Immunol **45**(7): 1863-71.

Van Gool, L. *et al.* (1997). "Overexpression of human poly(ADP-ribose) polymerase in transfected hamster cells leads to increased poly(ADP-ribosyl)ation and cellular sensitization to gamma irradiation." Eur J Biochem **244**(1): 15-20.

Virág, L. and C. Szabó (2002). "The therapeutic potential of poly(ADP-ribose) polymerase inhibitors." Pharmacol Rev **54**(3): 375-429.

Vodenicharov, M. D. *et al.* (2005). "Mechanism of early biphasic activation of poly(ADP-ribose) polymerase-1 in response to ultraviolet B radiation." J Cell Sci **118**(Pt 3): 589-99.

von Kobbe, C. *et al.* (2003). "Central role for the Werner syndrome protein/poly(ADP-ribose) polymerase 1 complex in the poly(ADP-ribosyl)ation pathway after DNA damage." Mol Cell Biol **23**(23): 8601-13.

von Kobbe, C. *et al.* (2004). "Poly(ADP-ribose) polymerase 1 regulates both the exonuclease and helicase activities of the Werner syndrome protein." Nucleic Acids Res **32**(13): 4003-14.

Wacker, D. A. *et al.* (2007). "The DNA binding and catalytic domains of poly(ADP-ribose) polymerase 1 cooperate in the regulation of chromatin structure and transcription." Mol Cell Biol **27**(21): 7475-85.

Wagner, S. *et al.* (2007). "Lactate down-regulates cellular poly(ADP-ribose) formation in cultured human skin fibroblasts." Eur J Clin Invest **37**: 134-139.

Wallace, D. C. (2001). "A mitochondrial paradigm for degenerative diseases and ageing." Novartis Found Symp **235**: 247-63; discussion 263-6.

Wang, M. *et al.* (2006). "PARP-1 and Ku compete for repair of DNA double strand breaks by distinct NHEJ pathways." Nucleic Acids Res **34**(21): 6170-82.

Wang, X. G. *et al.* (2007). "PARP1 Val762Ala polymorphism reduces enzymatic activity." Biochem Biophys Res Commun **354**(1): 122-6.

Wang, Z. Q. *et al.* (1995). "Mice lacking ADPRT and poly(ADP-ribosyl)ation develop normally but are susceptible to skin disease." Genes Dev **9**(5): 509-20.

Wang, Z. Q. *et al.* (1997). "PARP is important for genomic stability but dispensable in apoptosis." Genes Dev **11**(18): 2347-58.

References

Watson, A. J. et al. (1995). "Poly(adenosine diphosphate ribose) polymerase inhibition prevents necrosis induced by H2O2 but not apoptosis." Gastroenterology **109**(2): 472-82.

Wayman, N. et al. (2001). "5-aminoisoquinolinone, a potent inhibitor of poly (adenosine 5'-diphosphate ribose) polymerase, reduces myocardial infarct size." Eur J Pharmacol **430**(1): 93-100.

Weseler, A. R. et al. (2009). "Poly (ADP-ribose) polymerase-1-inhibiting flavonoids attenuate cytokine release in blood from male patients with chronic obstructive pulmonary disease or type 2 diabetes." J Nutr **139**(5): 952-7.

Whitacre, C. M. et al. (1995). "Involvement of NAD-poly(ADP-ribose) metabolism in p53 regulation and its consequences." Cancer Res **55**(17): 3697-701.

Whitaker, J. R. and P. E. Granum (1980). "An absolute method for protein determination based on difference in absorbance at 235 and 280 nm." Anal Biochem **109**(1): 156-9.

Wilson, G. L. et al. (1984). "Mechanisms of streptozotocin- and alloxan-induced damage in rat B cells." Diabetologia **27**(6): 587-91.

Woon, E. C. and M. D. Threadgill (2005). "Poly(ADP-ribose) polymerase inhibition - where now?" Curr Med Chem **12**(20): 2373-92.

Xiao, C. Y. et al. (2005). "Poly(ADP-Ribose) polymerase promotes cardiac remodeling, contractile failure, and translocation of apoptosis-inducing factor in a murine experimental model of aortic banding and heart failure." J Pharmacol Exp Ther **312**(3): 891-8.

Yamamoto, M. et al. (2009). "A multifunctional reporter mouse line for Cre- and FLP-dependent lineage analysis." Genesis **47**(2): 107-14.

Yano, K. et al. (2009). "Molecular mechanism of protein assembly on DNA double-strand breaks in the non-homologous end-joining pathway." J Radiat Res (Tokyo) **50**(2): 97-108.

Yelamos, J. et al. (2006). "PARP-2 deficiency affects the survival of CD4+CD8+ double-positive thymocytes." EMBO J **25**(18): 4350-60.

Yu, S. W. et al. (2006). "Apoptosis-inducing factor mediates poly(ADP-ribose) (PAR) polymer-induced cell death." Proc Natl Acad Sci U S A **103**(48): 18314-9.

Yu, S. W. et al. (2002). "Mediation of poly(ADP-ribose) polymerase-1-dependent cell death by apoptosis-inducing factor." Science **297**(5579): 259-63.

Zhang, S. et al. (2007). "c-Jun N-terminal kinase mediates hydrogen peroxide-induced cell death via sustained poly(ADP-ribose) polymerase-1 activation." Cell Death Differ **14**(5): 1001-10.

Zhang, X. et al. (2005). "Polymorphisms in DNA base excision repair genes ADPRT and XRCC1 and risk of lung cancer." Cancer Res **65**(3): 722-6.

Zingarelli, B. et al. (1997). "Protection against myocardial ischemia and reperfusion injury by 3-aminobenzamide, an inhibitor of poly (ADP-ribose) synthetase." Cardiovasc Res **36**(2): 205-15.

Zong, W. X. et al. (2004). "Alkylating DNA damage stimulates a regulated form of necrotic cell death." Genes Dev **18**(11): 1272-82.

7 Appendix

7.1 Abbreviations

aa	Amino acid
AIF	Apoptosis-inducing factor
BER	Base excision repair
cDNA	Complementary DNA
Cre	Cre recombinase
Cytg	Cytoglobin b gene
DBD	DNA-binding domain
Dex	Dexamethasone
h	Human
kDa	Kilo Dalton
m	Murine
MMS	Methyl methanesulfonate
MNNG	N-methyl-N´-nitro-N-nitrosoguanidine
MNU	N-nitroso-N-methylurea
NAD^+	Nicotinamide adenine dinucleotide
NER	Nucleotide excision repair
NF-κB	Nuclear factor kappaB
PAR	Poly(ADP-ribose)
PARG	Poly(ADP-ribose) glycohydrolase
PARP-1	Poly(ADP-ribose) polymerase-1
wt	Wild-type

Appendix

7.2 Figures

Figure 1: Schematic structure of the modular organization of human PARP-1. 9

Figure 2: Schematic illustration of poly(ADP-ribose) metabolism. 11

Figure 3: The role of PARP-1/-2 in DNA repair. 17

Figure 4: The role of PARP activity in cell survival and cell death in response to DNA damage intensity. 21

Figure 5: Chemical structures of 3-AB, PJ34, BYK204165 and BYK236864. 28

Figure 6: Generation of transgenic mice. 55

Figure 7: Immunofluorescence analysis of H_2O_2 induced PAR formation in cultured mouse embryonic fibroblasts (3T3) from *Parp-1$^{+/+}$* mice exposed to BYK204165. 62

Figure 8: Immunofluorescence analysis of H_2O_2 induced PAR formation in cultured mouse embryonic fibroblasts (3T3) from *Parp-1$^{+/+}$* mice exposed to BYK236864. 63

Figure 9: Immunofluorescence analysis of H_2O_2 induced PAR formation in cultured mouse embryonic fibroblasts (3T3) from *Parp-1$^{-/-}$* mice exposed to BYK236864 or BYK204165. 63

Figure 10: Survival of COMF10 cells. 65

Figure 11: Survival of COR4 cells. 66

Figure 12: Transfection efficiency of hPARP-1 expression in EL-4 cells. 67

Figure 13: Influence of transfection reagent jetPEI^{TM} on DNA repair 68

Figure 14: Dose-dependent induction of DNA strand breaks by X-irradiation in EL-4 cells. 69

Figure 15: Time course of DNA strand break repair in EL-4 cells. 70

Figure 16: DNA strand break repair in EL-4 cells overexpressing hPARP-1 protein. 71

Figure 17: Immunofluorescence analysis of PAR formation induced by X-irradiation in cultured EL-4 cells in the presence of PJ34. 72

Figure 18: Repair of DNA strand breaks in EL-4 cells after PARP inhibition. 73

Figure 19: Schematic illustration of the created new MCS for generation of the transgene pUCTE5. 74

Figure 20: Schematic representation of the plasmids for the transgene generation which was then used for DNA microinjection to generate transgenic mice. 77

Appendix

Figure 21: Immunofluorescence staining for detection of transgene function in transfected hamster CO60 cells. 78

Figure 22: Western blot analysis of hPARP-1 and Cre recombinase in transfected hamster CO60 cells. 79

Figure 23: Genotyping of 79 putative *hPARP-1* founder mice by real-time PCR analysis. 80

Figure 24: Analysis of founder mice by conventional PCR. 81

Figure 25: Western blot analysis of hPARP-1 and Cre recombinase expression in *lck-Cre* x *hPARP-1* transgenic mice. 81

Figure 26: Western blot analysis of PARP-1 expression in kidney and spleen from *hPARP-1* x *EIIa* transgenic mice. 82

Figure 27: Excision of the *Neo*/Stop sequence after mating of *hPARP-1* x *lck-Cre* tg mice. 83

Figure 28: Excision of the *Neo*/Stop sequence after mating of *hPARP-1* x *EIIa-Cre* tg mice. 84

Figure 29: Genotyping of 3 to 4 founder mice in different regions of the transgene. 85

Figure 30: Analysis of mRNA of *hPARP-1* transgenic mice. 86

Figure 31: Real-time PCR standard curve for the neomycin cassette of the *hPARP-1* transgene. 87

Figure 32: Copy number of *hPARP-1* in transgenic founder mice. 88

Figure 33: Schematic representation of PGKneotpAlox2. 104

7.3 Tables

Table 1: Chemicals and reagents 31

Table 2: Laboratory equipment 34

Table 3: Buffers and solutions 35

Table 4: Plasmids 37

Table 5: Oligonucleotides 38

Table 6: PCR conditions 40

Table 7: DNA and protein ladders 40

Table 8: Kits 41

Table 9: Enzymes 41

Appendix

Table 10: Restriction enzymes 42

Table 11: Polymerases 42

Table 12: Antibodies 43

Table 13: Cell lines 43

Table 14: Organisms 44

Table 15: Software 44

Table 16: SDS-PAGE separating/stacking gel 56

Table 17: hPARP-1 transfection efficiency in EL-4 cells under different experimental conditions . 67

i want morebooks!

Buy your books fast and straightforward online - at one of world's fastest growing online book stores! Environmentally sound due to Print-on-Demand technologies.

Buy your books online at
www.get-morebooks.com

Kaufen Sie Ihre Bücher schnell und unkompliziert online – auf einer der am schnellsten wachsenden Buchhandelsplattformen weltweit! Dank Print-On-Demand umwelt- und ressourcenschonend produziert.

Bücher schneller online kaufen
www.morebooks.de

VDM Verlagsservicegesellschaft mbH
Heinrich-Böcking-Str. 6-8
D - 66121 Saarbrücken

Telefon: +49 681 3720 174
Telefax: +49 681 3720 1749

info@vdm-vsg.de
www.vdm-vsg.de

Printed by Books on Demand GmbH, Norderstedt / Germany